Xenobot Chronicles

Secrets of Living Machines

W J Francis

Copyright © 2024 by W J Francis
All rights reserved. No part of this book may be reproduced, distributed, or transmitted in any form or by any means, including photocopying, recording, or other electronic or mechanical methods, without the prior written permission of the publisher, except in the case of brief quotations embodied in critical reviews and certain other noncommercial uses permitted by copyright law.

DEDICATION

To the brilliant minds and tireless hands shaping the future of technology. Your innovation, perseverance, and passion power the world we live in today. This book is a tribute to your dedication in making the impossible, possible.

Disclaimer: The information contained in this book is for educational and informational purposes only. It is not intended as medical advice and should not be relied upon as such. The author and publisher are not responsible for any adverse effects or consequences resulting from the use of any information, suggestions, or recommendations in this book

XENOBOT CHRONICLES unveils a thrilling journey through the frontiers of bioengineering and artificial life. When a group of renegade scientists stumbles upon a groundbreaking discovery—a new form of life they name "xenobots"

These xenobots are no ordinary organisms. Shaped from living cells, these programmable life forms can heal, transform, and even self-replicate.

Told through the eyes of a diverse cast—a young scientist haunted by her past, a whistleblower caught in the crossfire, and a former soldier seeking redemption—**XENOBOT CHRONICLES** asks us to confront the ethical limits of creation.

XENOBOT CHRONICLES offers readers a compelling and scientifically-backed exploration into the revolutionary field of xenobots.

Drawing on extensive studies, groundbreaking case studies, and contributions from top biologists and engineers, this book provides a comprehensive understanding of the possibilities and challenges of this cutting-edge technology.

References to peer-reviewed papers and expert opinions ensure a thoughtful and credible presentation of the subject, making it an essential read for those looking to delve into the future of biotechnology and artificial organisms.

CONTENTS

CHAPTER 1: THE BIRTH OF XENOBOTS 7

CHAPTER 2: THE MERGING OF BIOLOGY AND TECHNOLOGY 25

CHAPTER 3: FROM CELLS TO XENOBOTS: THE BUILDING BLOCKS 38

CHAPTER 4: COMPUTATIONAL DESIGN AND SIMULATION 52

CHAPTER 5: EVOLUTIONARY ALGORITHMS IN XENOBOT DESIGN 66

CHAPTER 6: XENOBOT LOCOMOTION AND MOBILITY 79

CHAPTER 7: SENSING AND PERCEPTION IN XENOBOTS 92

CHAPTER 8: XENOBOT BRAIN: NEURAL NETWORKS AND CONTROL 106

CHAPTER 9: XENOBOTS AS ENVIRONMENTAL EXPLORERS 119

CHAPTER 10: XENOBOTS IN MEDICINE AND HEALTHCARE 132

CHAPTER 11: ETHICAL AND MORAL CONSIDERATIONS 145

CHAPTER 12: XENOBOTS AND THE FUTURE OF BIOTECHNOLOGY 158

CHAPTER 13: COLLABORATIONS BETWEEN BIOLOGISTS AND ENGINEERS 172

CHAPTER 14: XENOBOT-ORGANISM INTERACTIONS 185

CHAPTER 15: XENOBOT ETHICS AND REGULATIONS 198

CHAPTER 16: XENOBOTS IN EDUCATION AND RESEARCH 212

CHAPTER 17: THE XENOBOT ARTISTIC REVOLUTION 225

CHAPTER 18: CHALLENGES IN XENOBOT AUTONOMY 236

CHAPTER 19: XENOBOTS AND SPACE EXPLORATION 249

CHAPTER 20: XENOBOTS AND NANOTECHNOLOGY 262

CHAPTER 21: XENOBOTS AND BIOENERGY PRODUCTION 276

CHAPTER 22: THE DARK SIDE OF XENOBOTS 289

CHAPTER 23: XENOBOTS IN AGRICULTURE AND ENVIRONMENTAL CLEANUP..302

CHAPTER 24: PUBLIC PERCEPTION AND XENOBOT ACCEPTANCE 315

CHAPTER 25: XENOBOTS BEYOND EARTH 329

CHAPTER 26: THE EVOLUTION OF XENOBOT DIVERSITY 342

CHAPTER 27: XENOBOTS AND NEUROLOGICAL RESEARCH 356

CHAPTER 28: THE UNCHARTED ETHICAL TERRITORY OF XENOBOTS 370

CHAPTER 29: XENOBOTS AND THE QUEST FOR ARTIFICIAL LIFE 383

CHAPTER 30: THE EVER-EVOLVING XENOBOT REVOLUTION 395

REFERENCES: 408

Chapter 1: The Birth of Xenobots

The story of Xenobots begins in an era when the possibilities of synthetic life were barely imagined beyond science fiction. Named after the African clawed frog Xenopus laevis, the cells of these frogs became the building blocks for what we now call "Xenobots," the world's first programmable, living machines.

Developed in 2020 by a team of scientists from Tufts University and the University of Vermont, these microscopic organisms combined biology and artificial intelligence in a truly revolutionary way. Xenobots were not just another lab invention; they marked a paradigm shift that stretched the very boundaries of what it means to be a living entity.

Xenobots are neither purely biological nor purely mechanical—they occupy an intermediate space. They are small, often less than a millimeter wide, but packed with potential. Their creation involved assembling living frog cells in entirely new configurations that could be programmed to accomplish specific tasks, such as movement, self-repair, or even the ability to carry tiny payloads.

The cells used for Xenobots were sourced from frog embryos, specifically from heart and skin cells. By combining these cells in a controlled environment, scientists discovered that they could create entirely new life forms that are capable of behaviors unimaginable in their natural state.

The birth of Xenobots began with an innovative approach that brought together evolutionary algorithms and biology. Using a powerful supercomputer, scientists simulated thousands of potential cell configurations in virtual space. Each configuration, or "blueprint," was tested to see how it might behave and function. Some arrangements moved effectively, others clustered together, and a few exhibited behaviors that suggested autonomous coordination.

The most promising virtual designs were then translated into real-life organisms by arranging frog cells under a microscope. Heart cells, which naturally contract, became the "motors" for movement, while skin cells formed a structural framework. What emerged from these efforts were clusters of living cells that could move in coordinated patterns, a living embodiment of their digital design.

One of the most astonishing aspects of Xenobots is their self-repair ability. Unlike most machines, which suffer wear and tear and require external maintenance, Xenobots are living systems with a natural tendency to heal.

If a Xenobot is cut or damaged, the cells gradually knit themselves back together, restoring functionality without any need for intervention. This regenerative quality points to a biological resilience that traditional machinery cannot replicate.

By tapping into this intrinsic characteristic, scientists envision future applications where Xenobots could perform tasks in environments hostile to human intervention, such as the ocean depths or within the human body, with minimal risk of failure.

In addition to their self-healing, Xenobots offer a level of environmental sustainability that is rare in robotics. Unlike metal-based robots that require extraction, processing, and disposal of heavy metals and rare-earth materials, Xenobots are fully biodegradable.

When their lifespan ends, they naturally decompose without leaving harmful residues. This ecological harmony positions them as a groundbreaking technology for green robotics, sparking ideas about their potential roles in environmental cleanup or medical applications where biocompatibility is crucial.

The developmental pathway of Xenobots also provides fascinating insights into cellular behavior. Cells that would normally grow into specific frog tissues and organs were instead encouraged to form entirely new structures and behaviors.

This malleability challenges long-held views about cell specialization and raises profound questions about the nature of life. Cells, when placed in novel environments, can cooperate and adopt new roles, suggesting that cellular behavior is not as rigidly determined as once thought. Xenobots demonstrated that cells have a capacity for plasticity, reconfiguring themselves to function in ways that defy their natural destinies.

The rise of Xenobots has opened up unprecedented opportunities in medicine. Imagine tiny, biodegradable robots that could travel through the human body, delivering targeted drugs to cancer cells or clearing away plaque from arteries without any surgical procedures. As these technologies evolve, they could offer a new frontier in personalized medicine, adapting to the unique biological landscape of each patient.

Beyond medical applications, researchers are exploring the idea of Xenobots as environmental stewards, capable of locating and removing microplastics from the ocean, or even detecting and neutralizing toxic chemicals in polluted areas.

By harnessing the intelligence of their cellular structure, Xenobots can be programmed for specific, beneficial tasks that address some of the pressing challenges facing our planet.

Despite their potential, the birth of Xenobots has sparked ethical and philosophical debates. These new life forms blur the line between living organisms and artificial creations, raising questions about the definition of life and our responsibilities as creators.

Should we see Xenobots as tools, as we might with traditional robots, or do they deserve a different consideration because they are made from living cells? Concerns have also been voiced regarding control and containment.

As with any emerging technology, there are fears that unforeseen consequences could arise if Xenobots were used irresponsibly or inappropriately. Scientists and ethicists are working together to establish guidelines and regulations to ensure that Xenobot research advances in a responsible and thoughtful way.

The story of Xenobots is still in its infancy, and as research advances, we are likely to see even more surprising applications and capabilities. The potential for Xenobots to evolve into more complex forms is a real possibility, especially as scientists learn more about how to harness and direct cellular behavior.

Some envision a future where these bio-bots could become highly specialized for complex tasks, perhaps even learning from their environments in ways that mimic basic intelligence.

These advances could lead to biohybrid entities capable of adapting and responding to real-time conditions, revolutionizing fields ranging from healthcare to environmental science.

At its core, the birth of Xenobots reflects humanity's growing capacity to design life itself. By combining the forces of biology, computation, and engineering, scientists have crafted living, programmable entities that push the limits of what was previously possible.

Xenobots symbolize a significant leap forward, not only in the field of robotics but in our understanding of life and intelligence. As we stand at the frontier of this new field, the story of Xenobots reminds us that life is not a fixed, rigid state but a spectrum of possibilities.

The birth of these tiny bio-bots marks a pivotal moment in scientific history, one that could reshape our relationship with technology, the environment, and life itself.

The journey of Xenobots is only beginning, and as the Xenobot Chronicles unfold, these remarkable creations promise to reshape our world in ways that are as exciting as they are mysterious.

We are on the threshold of a new age, one where life can be molded, directed, and harnessed in ways that defy both expectation and imagination.

The birth of Xenobots, a blend of science and ingenuity, stands as a testament to the boundless potential of human innovation.

Introduction to Xenobots, the World's First Living Robots

In 2020, scientists introduced a groundbreaking creation to the world—a living organism designed and engineered by humans, known as a xenobot. Named after the African clawed frog, Xenopus laevis, from which they derive, xenobots are neither traditional robots nor typical biological organisms.

Instead, they are a unique hybrid: living cells organized in a way that allows them to function like programmable entities.

These cellular creations are the world's first "living robots," blurring the boundaries between life as we traditionally understand it and the mechanical intelligence we associate with robotics.

The Science Behind Xenobots

At the core of xenobot creation is a technique that combines stem cell science, computational design, and micro-assembly. The process starts with stem cells, which are undifferentiated cells capable of developing into various specialized types.

In this case, scientists used stem cells taken from the frog Xenopus laevis, specifically selected due to their versatility and robustness in laboratory environments.

Using computational modeling, scientists design a "blueprint" for the structure and functions they want the xenobots to perform.

Advanced algorithms simulate thousands of possible configurations, determining the most efficient ways for cells to arrange themselves to achieve the desired actions—whether it be movement, object manipulation, or clustering together.

The chosen design is then implemented by hand-assembling the cells under a microscope, where they self-organize into the predetermined forms and begin to function in line with the programmed behavior.

Once assembled, these cellular clusters can crawl, spin, and move in a petri dish, demonstrating programmed behaviors without a brain, nervous system, or DNA modifications. They don't reproduce or carry out metabolism in the same way as traditional organisms.

Instead, their lifespan and function are determined by their cellular composition and external environment. For example, their movement is powered by contractions in their muscle cells, while the passive shape of other cells directs the pattern and direction of movement.

What Makes Xenobots Special?

Xenobots represent a remarkable intersection of biology and robotics, with several unique properties that distinguish them from other robots or organic systems.

First, xenobots are fully biodegradable. They don't leave any environmental waste, a stark contrast to traditional robots built from metal or plastic. Once their task is complete or they run out of energy, xenobots naturally decompose, leaving no harmful byproducts behind. This property is particularly appealing in applications where non-toxic and eco-friendly solutions are essential.

Second, unlike traditional machines, xenobots have self-repair capabilities. If cut or damaged, xenobots can "heal" by bringing cells back together, a process reminiscent of wound healing in multicellular organisms.

This unique trait could make them highly resilient in performing tasks where minor wear or damage might otherwise reduce functionality.

Lastly, xenobots operate at the microscale, which allows them to explore environments too small or complex for larger robots to navigate. Imagine exploring fine soil particles to detect toxins, moving through tiny blood vessels for drug delivery, or breaking down harmful microplastics in the ocean.

Their tiny size and biological nature make them ideal for tasks where precision and eco-friendliness are paramount.

Potential Applications of Xenobots

Xenobots may still be in the early stages of development, but scientists are already envisioning a range of practical applications that could harness their unique properties. One area of interest is in medical science, particularly for minimally invasive procedures.

In theory, xenobots could be deployed within the human body to perform precise tasks, such as targeted drug delivery to a tumor or removing plaque from arteries, where traditional surgical methods would be too risky or invasive.

Since xenobots are made from biological cells, they could potentially operate within the human body without causing immune reactions, as the cells could be derived from the patient's own tissue in the future.

Another promising application is environmental. Due to their biodegradability, xenobots could be used to target and break down pollutants in delicate ecosystems without leaving residual waste.

For instance, they might be engineered to bind with and remove microplastics from the ocean or to detect and neutralize chemical spills.

By creating xenobots programmed to search for and degrade specific toxins or plastics, we could reduce environmental contaminants in ways that are ecologically friendly and self-contained.

In addition to these immediate applications, xenobots could serve as a revolutionary platform for studying fundamental biological processes.

Because they can be designed and observed in controlled conditions, xenobots offer scientists a model to understand how simple cell clusters can form functional, self-organized systems.

This could shed light on early developmental processes in multicellular organisms or provide insights into how to guide cell behavior for regenerative medicine.

Ethical and Philosophical Questions

The advent of xenobots opens the door to several ethical and philosophical considerations. If xenobots are classified as "living organisms," then where do they fall on the moral spectrum of life? They are not sentient and do not experience pain, as they lack any nervous system, but the prospect of creating "living machines" raises important questions about the limits of biotechnology.

Should there be restrictions on what scientists are allowed to design? At what point does human manipulation of life cross into areas that are ethically gray?

Further, the potential for xenobots in biomedicine or the environment must be balanced with considerations about control and safety. Although xenobots are biodegradable, ensuring they only perform their intended functions and remain harmless in complex ecosystems is critical. Regulatory frameworks will need to adapt to these emerging technologies to ensure that the benefits are realized responsibly and ethically.

The Future of Xenobots

The creation of xenobots marks a transformative step in synthetic biology, yet it is just the beginning. Researchers continue to refine their techniques to make xenobots more complex and functional. With improvements in computational modeling, scientists hope to program xenobots with more advanced behaviors, such as collective motion or adaptive responses to environmental changes.

This could involve designing "swarms" of xenobots that work together to accomplish tasks that would be impossible for a single unit, opening up possibilities in fields like search-and-rescue missions or environmental cleanup.

In the not-so-distant future, we might see xenobots with more specialized forms and functions, capable of carrying out a broad range of precise and sensitive tasks. They could pave the way for more sophisticated forms of biohybrid robots, ultimately advancing both our understanding of life and the tools we use to interact with it.

Xenobots represent not only a scientific achievement but a philosophical leap. They force us to redefine our understanding of "living" and "machine," hinting at a future where life itself may become a tool for advancing human capability and environmental stewardship.

This emerging technology exemplifies the potential of synthetic biology to challenge our assumptions and reshape our world in profound ways.

Historical Context and Development Timeline of Xenobots

Xenobots represent a fascinating fusion of biology, robotics, and artificial intelligence, offering a glimpse into the future of programmable living organisms.

This technological marvel did not appear out of thin air but is rooted in decades of advancements across multiple scientific fields.

Understanding the timeline of xenobot development provides essential insight into the collaborative nature of this breakthrough, which combines evolutionary biology, bioengineering, and computer science.

The Foundations: From Mechanical Robots to Biological Machines (1940s–1990s)

The origins of xenobot research are traceable back to the early days of robotics in the 1940s and 1950s, a period characterized by the invention of mechanical robots that could perform simple, programmed tasks.

These early robots were far from the bioengineered entities that xenobots would eventually become, but they laid the groundwork for creating entities that could follow programmed commands.

Around this same time, scientists began exploring cell biology in unprecedented detail, driven by advancements in microscopy and cellular biology that enabled them to understand cellular processes and interactions.

Throughout the 1970s and 1980s, researchers expanded their understanding of both robotics and cellular control mechanisms.

Bioengineering research surged, especially in fields like synthetic biology, where scientists attempted to modify or create biological systems with specific functionalities.

These early explorations of synthetic life helped establish a foundation for the possibility of manipulating biological tissue in new ways.

Early Concepts and Biohybrid Exploration (1990s–2000s)

The 1990s marked a turning point with advancements in genetic engineering, as the completion of the Human Genome Project in 2003 sparked a surge of interest in reprogramming biological organisms.

Researchers in synthetic biology began experimenting with ways to engineer microorganisms to perform tasks, such as producing chemicals or cleaning up oil spills.

However, these engineered organisms were primarily single-celled and microscopic.

Meanwhile, robotics and artificial intelligence continued to advance, giving rise to early ideas about hybrid systems that combine biological components with artificial control.

Scientists began envisioning the possibility of combining living tissues with synthetic materials to create so-called "biohybrid" robots—an idea that would ultimately inspire the xenobot.

Advances in soft robotics, which explored flexible and adaptable materials for robotic systems, were particularly influential, as they demonstrated how living tissues might be used to create machines capable of complex and organic movements.

The Intersection of AI and Biology: Toward Programmable Life (2010s)

The 2010s saw rapid advances in artificial intelligence, particularly in machine learning, which allowed computers to identify patterns and optimize solutions to complex problems. These developments caught the attention of bioengineers, who began applying machine learning techniques to biological problems.

This decade also saw the rise of bioprinting, a revolutionary technique that allowed scientists to "print" living cells in specific patterns and configurations.

Bioprinting provided new control over cellular structure, a capability that would later prove essential in designing xenobots.

Around the same time, AI-driven approaches like neural networks started being used in biological and biomedical contexts. In 2018, researchers realized that machine learning algorithms could optimize the design of biological systems, providing viable pathways to "grow" desired shapes and functions.

Scientists could theoretically ask an AI system to create a biological organism with specific characteristics, bringing them a step closer to creating a programmable biological entity.

The Breakthrough: Xenobots Emerge (2019)

In 2019, a collaborative team from Tufts University and the University of Vermont made headlines by creating the world's first xenobots. Their goal was to harness the power of AI to create a new type of programmable biological machine, and they succeeded by using frog (Xenopus laevis) embryonic cells.

This achievement was the result of a two-pronged approach: the Tufts team focused on manipulating the biology of the frog cells, while the University of Vermont team developed an AI algorithm to design the most efficient biological configurations.

The researchers employed a machine learning algorithm to run thousands of simulations, seeking arrangements of frog cells that could produce purposeful, directed movement. The algorithm generated various configurations of cells based on desired tasks, and the team chose specific designs to "build" using frog skin and heart cells.

The skin cells provided structural support, while the heart cells contracted rhythmically, creating the locomotion necessary for movement. These small, cell-based machines, measuring less than a millimeter, were named xenobots after the frog species they originated from.

Xenobots were a groundbreaking innovation because they displayed the capacity to perform simple tasks, such as moving in a chosen direction, carrying small loads, and working together to achieve collective tasks.

Additionally, they were able to self-heal when cut, highlighting their robustness and potential for applications in medicine and environmental cleanup.

Refinement and Expansion: New Capabilities (2020s)

Following their debut, xenobots underwent various refinements to enhance their functionality and versatility. In 2021, researchers made significant strides by developing self-replicating xenobots, which could reproduce under certain conditions.

This marked an unprecedented advancement in synthetic biology, as the xenobots used their collective behavior to gather loose cells and create "offspring" versions of themselves. The discovery of self-replication opened up further ethical and practical discussions on the potential uses and limitations of xenobot technology.

Scientists have continued to improve xenobot design, using AI algorithms to produce more complex and efficient configurations. These advancements have led to xenobots capable of different movement patterns, responsiveness to environmental stimuli, and the ability to perform more sophisticated tasks, such as forming shapes and navigating obstacles.

The latest research focuses on enhancing xenobots' longevity, intelligence, and adaptability, positioning them as valuable tools for medical applications, environmental cleanup, and even therapeutic interventions within the body.

Future Prospects: What Lies Ahead

The development of xenobots has opened up numerous possibilities for their application, from targeted drug delivery to deep-sea exploration. However, this journey is far from over.

As researchers continue refining xenobot capabilities, there are discussions around the ethical implications, safety, and control measures needed to manage these programmable biological entities responsibly.

Future xenobot iterations may have the capacity to self-repair on a molecular level, interact with their environments intelligently, and perhaps even possess limited forms of memory.

The timeline of xenobot development is a testament to the power of interdisciplinary collaboration and the incredible potential of merging biology with artificial intelligence.

By tracing this evolution from early robotics and bioengineering experiments to the creation of programmable life forms, we gain a better understanding of how far science has come—and how it may yet transform our world.

As we continue to chart the course of xenobot development, we are likely to witness even more astonishing advancements that redefine the boundaries between living organisms and machines.

The Breakthrough Discovery of Biologists and Engineers

In the heart of a biophysics lab at Tufts University, a team of biologists and engineers achieved a milestone that blurred the line between biology and robotics. Their work, described as the creation of the world's first "living robots," has since sparked worldwide interest and imagination. These tiny, programmed organisms, known as xenobots, were named after the African clawed frog Xenopus laevis, from which they derive their stem cells.

Developed with the help of evolutionary algorithms, xenobots are capable of performing basic tasks such as moving in a predetermined direction, pushing small particles, and working together in groups. They represent a remarkable blend of biological tissue and computational design—a unique living technology.

This innovation could open doors to advances in biomedicine, environmental cleanup, and regenerative medicine.

Here, we explore the discovery that brought xenobots into the world, its key scientific principles, and the collaborative nature of this revolutionary breakthrough.

Merging Biology and Robotics: An Unexpected Alliance

The project was initiated with a bold vision to build biological machines using living cells instead of metal or plastic. Dr. Michael Levin, a biologist at Tufts University with expertise in bioelectricity and morphogenesis, partnered with Dr. Joshua Bongard, a roboticist from the University of Vermont known for his work with evolutionary algorithms.

By combining their expertise, the team explored the potential for using organic tissue in place of synthetic materials to create "biological robots."

Unlike traditional robots, which are made of metals and alloys, these biological machines are assembled from living cells, giving them unique capabilities and a remarkable adaptability that mechanical systems can't replicate.

The research aimed to create simple, programmable organisms capable of performing basic functions. By harnessing the inherent properties of cells, particularly their ability to self-organize and respond to environmental cues, the team hoped to open up new applications in medicine, environmental cleanup, and bioengineering.

Building Xenobots: The Role of Evolutionary Algorithms

Creating xenobots required a unique combination of cutting-edge tools and methodologies. To design the basic structure of these living machines, the researchers turned to evolutionary algorithms—a type of artificial intelligence that simulates natural selection to optimize solutions over multiple iterations.

These algorithms allowed the researchers to model thousands of possible xenobot designs in a virtual environment, identifying which shapes and cell configurations would work best for the intended tasks.

Each design was tested for functionality in a digital simulation, assessing how the biological components might interact, move, or perform desired actions. The software would run through generations of simulated trials, discarding ineffective configurations and "evolving" the designs that demonstrated the most promise.

Once an optimal design was identified, the team could replicate it in real life using frog stem cells.

Sculpting Life: Building Xenobots with Frog Cells

Creating a xenobot from scratch involved more than just clever programming; it required meticulous cellular assembly. To build these structures, the researchers used stem cells harvested from Xenopus laevis embryos, a species of frog with cells that are particularly robust and versatile.
These stem cells were then manually sculpted and assembled under a microscope, following the blueprints provided by the evolutionary algorithms.

The researchers carefully manipulated the cells into specific shapes, such as tiny, rounded organisms or forms with appendages that allowed for movement. The cells naturally adhered to one another, creating a stable structure, and once assembled, the xenobots displayed life-like behaviors. Muscle cells were arranged to enable contraction, while skin cells provided structural support, creating a simple but functional "body" that could propel itself in a controlled manner.

Emergent Behaviors: How Xenobots Learn and Adapt

One of the most intriguing aspects of xenobots is their ability to display emergent behaviors. Despite being composed of simple cells and lacking a nervous system, xenobots can perform tasks such as self-propulsion, object transport, and group coordination.

This is made possible by the inherent properties of biological cells, which respond to environmental stimuli in ways that are somewhat predictable yet adaptable.

For instance, xenobots can navigate around obstacles and adapt their paths to reach a goal, an ability typically associated with more complex organisms. Additionally, certain configurations of xenobots have been designed to work together, showing a rudimentary form of collective behavior.

This capacity for cooperation and adaptability makes xenobots especially promising for applications where traditional robots might struggle, such as navigating complex environments or handling fragile biological materials.

Potential Applications of Xenobot Technology

The xenobot discovery holds promise for numerous fields. In medicine, xenobots could potentially be programmed to deliver drugs precisely within the human body or target specific areas for tissue repair.

They might also assist in surgical procedures by navigating intricate parts of the body without causing damage to surrounding tissues.

Furthermore, xenobots could eventually be engineered to remove plaque in arteries, target cancer cells, or serve as biocompatible replacements for certain medical tools.

In environmental science, xenobots could help in pollution control by collecting microplastics or toxic particles from water sources.

Since they're biodegradable, xenobots have the added benefit of disintegrating after completing their tasks, leaving no trace in the ecosystem.

This application could revolutionize our approach to environmental cleanup, offering a more sustainable alternative to traditional waste-removal methods.

Ethical Considerations and Future Challenges

While xenobots present remarkable opportunities, their creation also raises ethical questions. Some scientists and ethicists are concerned about the implications of creating programmable organisms and whether such technology could lead to unintended consequences.

The potential for misuse is also a concern; xenobots could theoretically be engineered for harmful purposes if the technology falls into the wrong hands.

Researchers and policymakers alike are now grappling with the need to establish ethical guidelines and regulatory frameworks for the development and application of living machines.

From a scientific standpoint, numerous challenges remain. As of now, xenobots are limited to relatively simple tasks and have a limited lifespan. Extending their longevity and control over their behavior will require significant advances in cellular engineering and robotics.

Moreover, researchers must also find ways to prevent xenobots from becoming uncontrollable or causing harm in unintended ways.

A Glimpse into the Future of Living Machines

The creation of xenobots marks a pivotal moment in the history of science, one that may reshape our understanding of life, technology, and the boundaries between them.

By merging biological and mechanical principles, researchers have created a platform with nearly limitless possibilities.

Future advancements in xenobot design and control could unlock new avenues for medical treatments, environmental solutions, and perhaps even insights into the nature of intelligence and life itself.

As xenobot research continues, the world waits to see how this novel technology will evolve and what possibilities it may unlock. It is an exciting frontier that challenges traditional views of what life can be and suggests a future where biology and engineering are intertwined in unprecedented ways.

The discovery of xenobots not only represents a breakthrough in scientific understanding but also opens up a new chapter in the story of life—a chapter that may redefine what it means to be alive.

Chapter 2: The Merging of Biology and Technology

The convergence of biology and technology is redefining our understanding of life and its potential. This fusion has given birth to astonishing innovations such as xenobots—tiny, programmable biological organisms that represent the intersection of cellular biology and artificial intelligence.

In the not-so-distant past, biology and technology were viewed as distinct fields; one studied natural life, while the other focused on machines and computation. Today, however, advances in biotechnology, AI, and robotics are erasing these boundaries, creating tools and organisms that blend the capabilities of both.

The Origins of Biohybrid Systems

Biohybrid systems, which combine biological elements with synthetic ones, have been in development for several decades. Early examples included simple devices like pacemakers, which electrically stimulate heart muscles, and insulin pumps that maintain blood sugar levels in diabetics. But over time, research evolved from devices that merely support human biological functions to creating systems that can be guided and programmed, actively interacting with their environment in complex ways.

This journey required scientific breakthroughs in two main areas: biological engineering and computational power. Biologists had to learn how to grow and manipulate living tissues, while computer scientists and engineers developed algorithms capable of instructing and interacting with these tissues in meaningful ways.

Together, they laid the groundwork for biohybrid creations that possess a degree of autonomy and even the potential to self-heal—features that bridge the best of both biological and synthetic worlds.

Creating Xenobots: A Biological-Technical Breakthrough

The development of xenobots is one of the most promising examples of this merging of biology and technology. Xenobots, created from frog embryo cells (specifically, the African clawed frog Xenopus laevis), are tiny organisms designed through computational modeling and crafted in the lab.

Researchers first use powerful algorithms to simulate different cell configurations and evaluate potential designs that maximize specific desired actions, like movement or object manipulation.

Once a suitable design is selected, it's brought to life in the lab. Scientists extract and sculpt cells to form tiny biological machines capable of moving independently and even cooperating to accomplish tasks. These organisms are not traditional robots with metal parts and electronic circuits, nor are they natural organisms—they are biohybrids, embodying both synthetic and biological attributes in a single living structure.

The Role of Artificial Intelligence in Xenobot Design

AI plays an essential role in xenobot creation by assisting scientists in predicting how cells will interact and behave when arranged in different formations.

Algorithms powered by machine learning simulate thousands of configurations to identify which cell clusters could achieve specific goals, such as moving in a particular direction or carrying a tiny payload.

Without AI-driven simulations, such intricate biological designs would require years of trial and error in a lab. By guiding these decisions, AI shortens the development timeline and ensures that scientists can create xenobots with precision and purpose.

The use of algorithms in creating biological structures is a profound development, marking the point at which machines are not just analyzing data from biological systems but actively contributing to the creation of new, functional forms of life.

Applications of Xenobots and Biohybrid Technology

Xenobots hold promise for a wide range of applications, many of which could redefine fields from medicine to environmental science. Their potential for targeted drug delivery, for instance, could lead to treatments that are highly effective while reducing side effects by delivering medication directly to affected areas.

In environmental applications, xenobots might someday help clean up microplastics or hazardous waste by navigating and breaking down harmful substances.

One of the most remarkable potential uses of xenobots lies in their capacity for self-repair. When damaged, these tiny organisms can fuse together and restore some of their original functionality, a property borrowed from the regenerative abilities of their frog cell origins.

This capability means that xenobots could be deployed in environments where other forms of technology might degrade or break down, offering resilience that is typically found only in biological organisms.

Ethical Implications of Biohybrid Creations

While the merging of biology and technology opens doors to significant advancements, it also raises ethical questions. The creation of xenobots challenges our traditional understanding of life, leading to concerns about control, unintended consequences, and the moral implications of engineering living organisms.

Questions arise: What defines a life form, and at what point do we ascribe rights or responsibilities to these creations? Are xenobots akin to animals in need of ethical consideration, or do they fall into a unique category that warrants different guidelines?

There's also the question of containment and control. Biohybrid organisms designed to operate autonomously could, under unforeseen circumstances, act outside their intended purpose, especially in open ecosystems.

This highlights the need for strict regulatory frameworks that balance innovation with safety, ensuring that biohybrid creations are carefully managed and studied before deployment in sensitive environments.

The Future of the Biological-Technical Frontier

As biology and technology continue to intertwine, the field of xenobiology and biohybrid systems is expected to expand. Beyond xenobots, scientists are experimenting with programmable cells, synthetic genomes, and organisms that can survive extreme conditions.

These advancements could lead to revolutionary tools for space exploration, deep-sea missions, and medical treatments beyond current limitations.

Moreover, as AI becomes more sophisticated, its role in biological engineering will likely grow. Future algorithms may design more complex organisms capable of adapting, learning, or even evolving in response to their surroundings.

Such a development would represent a monumental shift, as scientists move from creating static biological systems to dynamic entities that can respond to changes in real-time.

Bridging the Biological and Digital Worlds

At its core, the merging of biology and technology is a story of convergence—a gradual blending of disciplines that were once worlds apart.

This collaboration is allowing us to build a new class of living machines that may not only assist humanity but fundamentally change the way we interact with our world.

Xenobots are only the beginning; they offer a glimpse of a future where biology and technology merge seamlessly, creating solutions that are as resilient, adaptable, and efficient as the natural systems that inspire them.

In envisioning this future, we are invited to reconsider the relationship between human ingenuity and the natural world. With the power to shape life at a cellular level, we stand on the precipice of a new era, where technology does not simply mimic life but actually becomes life in its own right.

The implications of this transformation are profound, as we look toward a future where the boundaries of biology and technology dissolve, unlocking possibilities we are only beginning to understand.

Exploring the Intersection of Biology and Robotics

The concept of merging biology and robotics might seem straight out of science fiction, but it's becoming an exciting scientific reality. Imagine living, functional machines made from biological tissue—machines that can move, heal, and adapt like living organisms.

This is the foundation of xenobots, tiny "biobots" created from living cells. Xenobots are the result of interdisciplinary science, bringing together knowledge from biology, robotics, artificial intelligence, and regenerative medicine.

They illustrate the potential of "soft robotics," a subfield of robotics focusing on adaptable, flexible machines built from soft, often biological materials.

Xenobots get their name from Xenopus laevis, the African clawed frog, whose cells form the building blocks of these unique living machines.

Using frog cells was an early breakthrough because of their regenerative capabilities, flexibility, and ability to be reprogrammed to serve new purposes outside their normal biological role.

By arranging these cells in specific structures and programming them with genetic instructions or external cues, scientists can guide xenobots to perform various tasks.

These could range from moving towards a target, collecting microscopic particles, or even self-organizing into clusters.

The creation of xenobots is a step forward for robotics because it taps into the natural complexity of biology. Traditional robots are built from non-living materials, such as metals and plastics, and rely on batteries, motors, and programming. By contrast, xenobots are composed of living cells, which already possess the internal machinery to move, process energy, and even repair themselves.

This shift from mechanical to biological design allows xenobots to operate in ways that were previously unimaginable. For example, rather than wearing down over time, xenobots can regenerate damaged parts or even reshape themselves if needed. Such abilities give xenobots a potential advantage in tasks that involve navigating complex or unpredictable environments.

At the heart of xenobot creation is a concept known as "biological design." This process involves arranging cells in such a way that they collectively exhibit behaviors beyond what individual cells could accomplish.

Using techniques like computer modeling, scientists simulate various configurations of cells to see how they might interact in a specific arrangement. Once they find a promising design, they create it in a lab using microsurgery or cell manipulation techniques.

For example, scientists might place muscle cells in specific locations to create movement or include skin-like cells to form protective layers. When these cells work together, they become something entirely new: a programmable living organism with capabilities that go beyond what any single cell type could achieve alone.

One of the fascinating aspects of xenobots is how they blur the line between living organisms and robots. Because they're made from cells, they share certain characteristics with living things—they can grow, change shape, and respond to their environment. But unlike traditional organisms, xenobots lack a central nervous system, organs, or even a brain.

Their behavior is driven by the basic, natural properties of cells, such as their ability to contract, move, and communicate with neighboring cells.

This makes xenobots an intriguing middle ground between biological life and machines, offering scientists insights into how life functions at a cellular level, while also providing a platform for new types of robotic applications.

Xenobots are also advancing thanks to artificial intelligence (AI).

Machine learning algorithms play a crucial role in designing xenobots by running simulations that predict how different cell arrangements might behave. In these simulations, the computer can run through countless possible configurations, testing each to see if it achieves a desired outcome—like moving in a specific direction or clustering in response to certain stimuli. Once a promising design emerges, researchers replicate it using real cells in a laboratory setting.

This AI-driven process accelerates xenobot development, allowing scientists to experiment with novel designs faster than would be possible through traditional trial and error.

As xenobot technology advances, potential applications are beginning to take shape. One promising area is environmental cleanup. Since xenobots can navigate aquatic environments and collect small particles, they could one day be deployed to clean up microplastics or other pollutants from water sources.

Unlike conventional robots, xenobots can operate in sensitive ecosystems without adding to pollution or leaving behind non-biodegradable waste, as they naturally break down after a certain period.

Researchers envision future xenobots that could seek out and remove toxins from contaminated sites, making them effective tools for environmental health.

In medicine, xenobots hold the potential to revolutionize treatments at the cellular level. For instance, they could be designed to deliver targeted therapies, removing damaged cells or pathogens without affecting healthy tissue. Due to their small size and biocompatibility, xenobots could safely operate within the human body, accessing hard-to-reach areas and providing localized treatments that reduce side effects.

Imagine a xenobot delivering a medication directly to an infected cell or carrying stem cells to a site that requires regeneration.

These applications are still in early development, but the possibilities are inspiring scientists to explore further.

While xenobots represent an incredible leap forward, they also raise important ethical and scientific questions. Since xenobots are technically alive, do they have rights or ethical considerations that apply to other living organisms? Should there be limits on how much autonomy we give them, or where they are deployed? And if xenobots evolve, could they develop behaviors that were not originally programmed?

These questions highlight the need for careful consideration as xenobot technology progresses. Scientists, ethicists, and policymakers will need to collaborate to ensure that xenobots are used responsibly and safely.

The intersection of biology and robotics, as illustrated by xenobots, is an evolving frontier that brings together the organic and the synthetic in unprecedented ways.

Unlike traditional robots, xenobots offer a more flexible, adaptive approach to robotics, showing how we can harness nature's own tools to solve complex problems.

While the science behind xenobots is still young, it offers glimpses of a future where the boundary between machines and living systems becomes increasingly blurred.

From cleaning polluted waters to delivering precision treatments, xenobots embody a unique blend of biological adaptability and technological ingenuity.

As research continues, we can expect even more groundbreaking applications and, along with them, a new era of scientific and ethical exploration.

Bioengineering and Its Applications in Creating Xenobots

In recent years, bioengineering has taken tremendous strides forward, redefining our understanding of biology and technology, particularly in the field of xenobot creation. Xenobots, named after the African clawed frog Xenopus laevis from which they derive, are a pioneering class of living, programmable organisms designed to perform a variety of tasks at a cellular level.

These biological robots, or "xenobots," blend biology and robotics in ways that could have broad implications across medicine, environmental science, and bioengineering.

Understanding Bioengineering's Role in Xenobot Development

Bioengineering is an interdisciplinary field that applies principles from biology, physics, chemistry, and engineering to create new forms of biological life or modify existing ones for innovative purposes. At the heart of xenobot creation lies the fusion of computer science, developmental biology, and robotics, facilitated by bioengineering techniques.

Researchers work at the cellular level, using stem cells and advanced algorithms to design these tiny, programmable entities. This process involves isolating cells, stimulating specific growth patterns, and programming them to perform desired behaviors or functions.

The production of xenobots begins with extracting cells from the Xenopus laevis frog and reconfiguring them into forms that mimic traditional robotics parts, such as appendages for movement. Using computer simulations, bioengineers test various designs that could be executed by clusters of these reprogrammed cells. After extensive simulations and modeling, they then proceed with actual cell manipulation and assembly, leading to the creation of the first functional xenobots.

This integration of computer modeling with live cellular components is revolutionary, representing a significant step forward in biotechnology and synthetic biology.

Applications of Xenobots in Medicine

One of the most exciting potential applications of xenobots is in the field of medicine. Xenobots could be programmed to perform microscopic surgical procedures, navigate complex biological environments, and deliver targeted drug therapies to specific tissues or cells in the human body.

For instance, xenobots may one day be capable of identifying cancerous cells and delivering chemotherapy directly to tumors, reducing the harmful side effects of traditional chemotherapy. They might also perform internal wound healing by navigating to damaged tissues and releasing growth factors or repairing cells directly.

Another promising application lies in their potential to clear out arterial plaque, which can cause heart disease. The development of bioengineered xenobots could lead to "micro-cleaners" that patrol the vascular system, dissolving dangerous plaque buildup in arteries and preventing strokes and heart attacks.

Xenobots, being soft and biodegradable, would have an advantage over traditional metal or plastic-based microsurgeries since they are less likely to cause immune reactions and can decompose naturally after fulfilling their tasks.

Environmental Applications: Xenobots as Bio-Remediators

Beyond medicine, xenobots hold significant promise in environmental science, especially for applications in bio-remediation—the use of organisms to clean up environmental contaminants. Xenobots could be deployed to bodies of water, where they would identify and break down pollutants such as microplastics or toxic chemicals, which are challenging to remove with conventional methods.

Their organic composition and small size allow them to move through ecosystems in a way that is minimally invasive, reducing environmental disturbance while improving pollution control. They can also be programmed to target specific pollutants, making them potentially efficient and selective agents in environmental cleanup efforts.

Another advantage of xenobots in environmental applications is their ability to function autonomously.

Once programmed, they could be released in polluted areas and left to operate independently, using basic stimuli-response systems to navigate their environment.

If successfully scaled, xenobot technology could be instrumental in addressing large-scale pollution issues that are difficult or costly to tackle with traditional methods, providing a sustainable and self-regulating solution.

Bioengineering Xenobots for Future Scientific Exploration

Xenobots represent not only a tool for practical applications but also a unique means of studying biological development and cellular behavior.

By engineering cellular robots, scientists are better able to understand how cells can work cooperatively to create complex systems—a phenomenon observed in organ and tissue formation.

Xenobots provide a controllable platform to explore how various cell types interact, communicate, and self-organize, which could lead to insights into tissue regeneration, organ growth, and even fundamental aspects of life itself.

This ability to bioengineer controlled biological entities opens new pathways for studying diseases at a cellular level, potentially leading to better therapeutic interventions and diagnostics.

For example, by engineering xenobots with disease-like properties, scientists can observe the early stages of diseases like cancer, Alzheimer's, or autoimmune disorders in a controlled setting, providing valuable information about these conditions' progressions and cellular effects.

Bioethics and the Future of Xenobot Applications

With these exciting advancements in bioengineering and xenobot technology comes an array of ethical considerations. As with any powerful technology, it's essential to consider the potential risks and societal implications of xenobots.

Questions around safety, environmental impact, and the potential for misuse must be addressed. Researchers and ethicists are working to establish guidelines and ethical frameworks to ensure that xenobot technology is used responsibly and sustainably.

Another important question is how xenobots might impact biodiversity. While they can help remediate environmental issues, their introduction to ecosystems must be carefully managed to prevent unintended consequences.

Xenobots are engineered to decompose naturally, but large-scale deployment or improper programming could lead to unforeseen ecological effects.

The Road Ahead for Bioengineering Xenobots

As bioengineering continues to evolve, the design and functionality of xenobots will only improve.

Current xenobots are relatively simple in their tasks, but as scientists learn more about programming cells to perform complex functions, the capabilities of xenobots will expand.

Future generations may feature enhanced mobility, communication between xenobot units, or the ability to conduct advanced problem-solving tasks in response to specific environmental cues.

The creation of xenobots has ushered in a new era for bioengineering, setting the stage for advanced, biologically-based machines that could complement or replace traditional robotic and medical technologies.

From performing precise surgeries to cleaning up oceans, xenobots are a testament to what can be achieved when biology and technology merge.

These bioengineered organisms are still in their infancy, yet they hold enormous promise, inspiring the next wave of breakthroughs in biotechnology and synthetic biology.

Bioengineering's role in developing xenobots reflects a pivotal point in our relationship with technology, shifting from mechanical to organic and programmable forms that can perform tasks in ways we are only beginning to understand.

With continued innovation and responsible development, xenobots may become a cornerstone of medical, environmental, and scientific solutions in the near future.

Chapter 3: From Cells to Xenobots: The Building Blocks

Imagine a world where the boundary between living organisms and engineered machines blurs, a world where biological tissues can be programmed to perform functions typically reserved for robots. This is the world of xenobots, a breakthrough in the merging of biology with robotics. These tiny, self-assembling creatures, created from living cells, are pioneering a new era in science.

Xenobots are not artificial in the traditional sense; they are built from real, living cells that have been reconfigured to perform specific tasks. But how do scientists go from single cells, the building blocks of life, to a functioning xenobot? This journey begins with a deep understanding of cellular biology and the art of cellular assembly.

1. Understanding the Cell: The Smallest Unit of Life

Cells are the fundamental building blocks of all living things. They are the smallest unit of life capable of performing essential functions such as growth, reproduction, and energy conversion. Each cell is a complex system with various specialized parts, or organelles, that contribute to its function.

For example, the nucleus contains the cell's DNA, which directs cellular activities and provides the genetic blueprint for life. The mitochondria generate energy, acting as the cell's powerhouse, while the cell membrane regulates interactions with the external environment.

Cells come in many different types, each adapted to perform specific roles. In the human body alone, there are hundreds of different cell types, including muscle cells, nerve cells, and skin cells. The versatility of cells is what makes xenobot creation possible. Scientists start with simple frog embryo cells, which have the potential to grow into various tissues. By repurposing these cells, they can be guided to form entirely new structures with new capabilities.

2. Stem Cells: The Key to Xenobot Development

At the core of xenobot creation are stem cells. Stem cells are unique because they have the potential to differentiate, or transform, into almost any cell type in the body. This makes them highly valuable for regenerative medicine, tissue engineering, and now, xenobot development.

When creating a xenobot, scientists often use frog stem cells from the species Xenopus laevis, which is why they are called xenobots. These cells can be harvested from frog embryos and cultured in a laboratory setting.

Through careful manipulation, scientists can guide these stem cells to aggregate and differentiate in controlled ways. By adjusting environmental conditions such as temperature, nutrient supply, and chemical signals, they can steer these cells to form clusters that exhibit coordinated behaviors.

Essentially, they provide the cells with an environment that encourages them to work together, leading to the formation of a xenobot.

3. Cell Aggregation and the Emergence of Collective Behavior

A unique aspect of xenobots is how the cells, when aggregated, display behaviors that single cells cannot achieve alone. When individual cells come together, they can communicate and coordinate their actions through chemical and electrical signals.

This collective behavior is an emergent property, meaning that it arises from the interactions of individual cells, leading to new, unexpected characteristics.

For example, muscle cells can be added to a xenobot to enable movement. These cells are known for their contractile ability, which they use to generate force. By strategically placing muscle cells within the structure, scientists can create xenobots that "walk" or swim in specific directions.

The addition of sensory cells can make xenobots responsive to certain stimuli, such as light or temperature. The arrangement and type of cells determine how the xenobot will function, much like how the parts of a machine determine what it can do.

4. The Role of Computer Modeling and AI in Xenobot Design

To design complex xenobots, scientists turn to computer modeling and artificial intelligence. Computer models allow researchers to simulate different cellular arrangements and test how they might behave without having to create each prototype in the lab.

AI algorithms can analyze vast amounts of data to predict which combinations of cells and structures will yield the most effective xenobot designs.

For instance, an AI algorithm might simulate thousands of potential xenobot structures and evaluate their efficiency for a particular task, such as transporting particles. Through this process, scientists can select the most promising designs and then attempt to recreate them with living cells in the lab.

This combination of AI and cellular biology allows for rapid experimentation and optimization, which would be much slower and more labor-intensive without computational assistance.

5. Constructing a Xenobot: The Physical Assembly Process

The physical creation of a xenobot requires delicate manipulation of cells under a microscope. Scientists use tiny, precise tools to assemble clusters of cells into specific shapes.

In many cases, the cells are encouraged to self-assemble based on cues from their environment. This is possible because cells naturally adhere to each other and can be directed to form organized structures.

Once the desired structure is achieved, the xenobot is placed in a petri dish to test its functionality. Researchers observe its movement, responsiveness, and durability, making adjustments as necessary.

This iterative process allows for the refinement of xenobot designs, ensuring that they are capable of performing the intended tasks.

6. Functions and Capabilities of Xenobots

The capabilities of xenobots are both fascinating and somewhat surprising, given their simplicity. Although xenobots lack a nervous system or brain, they can perform basic tasks such as locomotion and even limited self-repair.

For instance, if a xenobot is cut in half, it may be able to heal itself and continue moving. This self-healing property is due to the innate regenerative capabilities of the cells used to construct them.

In addition to movement, xenobots can be engineered to carry small particles, making them potential vehicles for delivering medication within the human body or transporting harmful waste in aquatic environments.

Researchers have even demonstrated that xenobots can "scoop up" microscopic particles, making them useful for environmental cleanup.

7. The Future Potential of Xenobots

Xenobots represent only the beginning of what might be achieved by combining biology with robotics. As scientists gain more control over cellular behavior and advance computer modeling techniques, the range of applications for xenobots is expected to expand.

In medicine, xenobots could be used for targeted drug delivery, navigating through blood vessels to reach specific areas of the body with precision. They may also play a role in regenerative medicine by helping repair damaged tissues or organs.

In environmental science, xenobots could be deployed in water systems to monitor and clean pollutants.

Because xenobots are biodegradable, they could offer a safer alternative to traditional micro-robots, which are typically made from non-degradable materials.

8. Ethical Considerations and Future Challenges

Despite the promising potential of xenobots, their development raises important ethical questions. As creatures that blur the line between life and machine, xenobots challenge our understanding of life itself.

How do we categorize these tiny living machines, and what ethical guidelines should govern their use?

Moreover, as xenobot technology advances, there are questions about potential risks, such as unintended ecological impacts or misuse for harmful purposes.

In summary, xenobots are an extraordinary innovation that bridges biology and technology. Created from living cells and guided by the principles of cellular biology, computer modeling, and artificial intelligence, xenobots illustrate the vast potential of engineered life.

From the basic unit of the cell to a functioning organism, the building blocks of xenobots represent a new frontier in science with the power to revolutionize medicine, environmental science, and perhaps our very understanding of life.

Understanding Stem Cells and Their Role in Xenobot Creation

Xenobots represent a fascinating frontier at the intersection of biology and robotics, taking their name from the African clawed frog, Xenopus laevis, whose cells serve as their fundamental building blocks.

The creation of these tiny, living robots is a feat achieved by harnessing the unique potential of stem cells.

Understanding what stem cells are, how they function, and why they are so critical in the assembly of xenobots reveals a lot about the future of synthetic biology and regenerative medicine.

What are Stem Cells?

Stem cells are often referred to as the "building blocks" of life. Unlike specialized cells such as neurons or muscle cells, stem cells are unspecialized, meaning they have the extraordinary ability to develop into many different cell types. This potential to differentiate into various cells underpins much of their promise in medicine.

Stem cells are categorized into types based on their potential:

1. Totipotent Stem Cells: These cells have the potential to develop into any cell type in an organism, including cells that make up extraembryonic structures like the placenta. In mammals, totipotency is limited to cells present in the very earliest stages after fertilization.

2. Pluripotent Stem Cells: Found within the embryo, these cells can differentiate into nearly any cell in the body. Embryonic stem cells are a prime example of pluripotent cells.

3. Multipotent Stem Cells: These cells can differentiate into a more limited range of cells. For example, hematopoietic stem cells found in bone marrow can turn into various types of blood cells but not into brain or muscle cells.

Stem cells' ability to transform into other types of cells gives scientists a powerful tool for engineering living structures—such as xenobots—capable of autonomous functions like movement and repair.

The Role of Frog Stem Cells in Xenobot Creation

To create xenobots, scientists turned to the stem cells of Xenopus laevis embryos. These frog embryos provide a source of cells that, under the right conditions, can be coaxed into forming new structures. But what makes these frog stem cells particularly valuable?

1. Flexibility and Adaptability: Xenopus stem cells, though multipotent, display remarkable adaptability. When removed from their native embryonic environment and exposed to new conditions, they can be guided to form simple living robots.

2. Non-genetic Programming: A key aspect of xenobot creation is that it doesn't involve altering the frog's DNA. Instead, these cells are manipulated mechanically and environmentally to assemble into new shapes and take on functions they wouldn't normally adopt.

This manipulation allows the cells to cooperate in a coordinated structure without requiring genetic modification.

3. Structural Integrity: Frog cells provide a sturdy foundation for creating functional shapes. When reconfigured into specific geometries, these cells can act collectively, giving rise to motile organisms that can navigate through their environment, perform basic tasks, and even exhibit basic forms of "memory" in their behavior.

How Stem Cells are Assembled into Xenobots

The process of assembling stem cells into xenobots relies on a delicate balance of biological science and engineering.

Unlike traditional robots made of metal or plastic, xenobots are crafted by sculpting clusters of living cells into specified designs.

Here's a breakdown of the process:

1. Isolation of Cells: Scientists extract skin and cardiac cells from Xenopus embryos. Skin cells act as a structural frame, while cardiac cells provide motility through their natural contractile behavior.

2. Designing Structures with Computer Models: Before assembling the cells, researchers use computer simulations to model various configurations and test the potential outcomes.

This process allows them to predict how certain designs will affect movement, shape retention, and overall functionality.

Advanced algorithms simulate how different arrangements of cells might interact, revealing the optimal structure for achieving desired behaviors.

3. Manual Assembly of Cells: Once a suitable configuration is selected, the cells are manually placed in arrangements predicted by the computer models. Using precise tools like micro-tweezers, scientists sculpt clusters of cells into predetermined shapes. For example, a simple spherical structure allows for forward movement, while a U-shaped design might enable more complex motions.

4. Emergence of Behavior: The assembled cells begin to behave as a coherent unit. Cardiac cells' rhythmic contractions generate movement, while skin cells hold the form. The resulting xenobot may then move through its environment, responding to stimuli such as chemicals or other cells, which enables it to perform simple tasks.

How Xenobots Utilize Stem Cells to Achieve Functionality

Xenobots, although rudimentary compared to more sophisticated forms of life, demonstrate surprising versatility.

This functionality is largely due to the inherent properties of stem cells:

1. Self-Repair: Stem cells in xenobots enable a limited degree of self-repair. If damaged, xenobots can sometimes reorganize their cells to restore function. This potential for self-repair hints at future applications in fields where self-healing materials are desirable.

2. Environmental Sensitivity: Xenobots respond to their surroundings, a quality largely attributed to the cells' natural ability to respond to biochemical signals. For instance, they can "sense" and swim towards or away from chemical gradients, a useful trait for potential applications in environmental cleanup or targeted drug delivery.

3. Collective Motion: The cells in xenobots work together, exhibiting a collective behavior that is not encoded by DNA but is a result of the way cells interact.

The mechanical interplay among cells—especially between skin and cardiac cells—allows xenobots to function as unified, albeit simple, organisms.

Potential Applications of Stem Cell-Based Xenobots

The discovery of xenobots opens exciting new paths for research. Since they are biodegradable, xenobots could offer a safer alternative to synthetic robots in certain applications.

Potential uses include:

1. **Environmental Remediation:** Due to their small size and biodegradability, xenobots could be released into ecosystems to perform targeted cleanup tasks, like collecting microplastics or breaking down pollutants.

2. **Targeted Drug Delivery:** Xenobots' capacity to move through environments and respond to biochemical cues means they could one day be programmed to deliver drugs directly to targeted sites in the human body, reducing side effects and improving efficacy.

3. **Regenerative Medicine:** Understanding how cells can be guided to take on new forms and functions could revolutionize tissue engineering.

Xenobots offer insights into cellular organization and could pave the way for constructing tissues, organs, or even fully functional biological machines.

Ethical Considerations

With these remarkable capabilities, xenobot research raises significant ethical questions.

Although xenobots are not genetically modified or engineered to have consciousness, their development touches on debates about artificial life and the manipulation of living cells.

As with any new technology, careful regulation and ethical guidelines will be essential to ensure that xenobot applications are safe, environmentally sound, and used responsibly.

The creation of xenobots is a pioneering achievement that underscores the vast potential of stem cells.

By leveraging the adaptability and natural plasticity of frog stem cells, scientists have unlocked new possibilities for building living machines.

As we continue to explore this fascinating field, xenobots may one day serve as tools for ecological repair, medicine, and a deeper understanding of life itself.

Cell Specialization and Differentiation

Imagine a world where every cell in our body—whether it's a muscle, nerve, or skin cell—was the same. We'd be like a disorganized mass, unable to perform the complex functions that make up life.

Instead, our bodies are beautifully orchestrated systems, each type of cell playing a unique role.

This specialization is no accident; it's the result of a remarkable process called cell differentiation, where unspecialized cells, or stem cells, transform into specific cell types, each with its own structure and function.

The Foundations of Cell Specialization

Cell differentiation is essential for building complex organisms, and it starts early in development. Initially, a fertilized egg is a single cell containing all the genetic information needed to make every type of cell in the body.

As this cell divides, the resulting cells form a ball called a blastocyst, which includes cells that can still become many different types.

This stage contains what scientists call pluripotent stem cells, which have the potential to develop into almost any cell type.

As cells continue dividing, they receive signals—sometimes from other cells or from their environment—that tell them to activate or suppress certain genes. Genes are segments of DNA that serve as instructions for making proteins, which drive cellular function.

When a gene is turned "on" or "off," it influences the proteins that a cell produces, and over time, this shapes the cell's function. Through this process, cells gradually take on specific roles, such as becoming muscle cells to help us move, or neurons to transmit signals in our nervous system.

How Differentiation Happens: The Role of Gene Expression

The journey from stem cell to specialized cell is largely guided by gene expression. This term refers to how active certain genes are in a cell at any given time. Every cell in our body has the same DNA, yet different types of cells express different sets of genes. For example, heart muscle cells and liver cells have the same DNA, but they function differently because they activate different genes.

Gene expression is controlled by transcription factors, which are proteins that bind to DNA and either promote or block the transcription of specific genes. When a cell encounters a particular transcription factor, it signals which genes to activate or deactivate. If a cell is destined to become a neuron, it will express genes that enable it to transmit electrical signals, while suppressing genes meant for other functions, like contracting muscles or producing digestive enzymes.

The Stages of Cell Differentiation: From Stem Cells to Specialized Cells

Differentiation happens in a few key stages, and each step represents a cell becoming more committed to a particular role:

1. Totipotency: The earliest cells in a developing embryo are totipotent, meaning they can give rise to any cell type in the body or even form an entire organism. For example, a fertilized egg is totipotent.

2. Pluripotency: As cells divide, they become pluripotent, able to differentiate into nearly any cell type in the body but not into an entire organism.

These are the cells in the inner cell mass of the blastocyst.

3. Multipotency: As development continues, cells become multipotent, meaning they can become a limited range of cells within a specific tissue or organ.

Blood stem cells, for example, are multipotent—they can become red blood cells, white blood cells, or platelets, but not skin or muscle cells.

4. Unipotency: Finally, cells reach a stage called unipotency, where they are fully specialized and can only divide to produce more of the same cell type. Skin cells, for example, are unipotent, only able to generate other skin cells.

Signals and Pathways That Guide Differentiation

Cells are constantly communicating with each other through biochemical signals, guiding the differentiation process.

These signals, often in the form of molecules like proteins or hormones, bind to receptors on cell surfaces, triggering a cascade of internal reactions that influence gene expression.

Some of the key pathways include:

• **Wnt Signaling Pathway:** This pathway is crucial in determining cell fate and plays a significant role in embryonic development. Depending on the levels and timing of Wnt signals, cells may turn into muscle, nerve, or connective tissue cells.

• **Notch Signaling Pathway:** Notch signaling helps maintain balance in cell differentiation and is vital in forming tissues and organs. This pathway is particularly important in stem cells within adults, helping tissues regenerate by producing new specialized cells.

- **Growth Factors:** Various proteins, such as epidermal growth factor (EGF) and fibroblast growth factor (FGF), act as signals that tell stem cells what type of cells they should become. These signals are commonly used in laboratories to guide stem cell differentiation for regenerative medicine.

Cell Specialization in Action: From Laboratory to Xenobots

Xenobots, tiny biological robots made from frog cells, showcase the principles of cell differentiation in a fascinating way. Scientists use stem cells from the African clawed frog (Xenopus laevis) and direct them to become specific types of cells, like skin or muscle cells.

These cells then self-organize into simple, functional robots that can move, work together, and even repair themselves.

When creating xenobots, scientists exploit the natural tendencies of muscle cells to contract and epithelial cells (a type of skin cell) to provide structural support.

By arranging these differentiated cells in particular patterns, researchers enable xenobots to perform basic tasks, such as swimming or carrying small objects.

This breakthrough demonstrates how understanding and controlling cell differentiation can lead to exciting applications in robotics and medicine.

Differentiation and Its Role in Regeneration and Healing

Our bodies are equipped with certain adult stem cells, which remain undifferentiated throughout life and aid in regeneration and repair.

These cells can divide to replenish specific types of cells when tissues are damaged.

For instance, muscle stem cells can replace muscle tissue lost to injury, while skin stem cells continually replace skin cells that die off.

Differentiation in regenerative medicine aims to harness this natural ability to repair tissues and even grow new organs. Scientists use stem cells to produce heart cells, neurons, and other types, hoping to treat diseases such as heart failure and neurodegenerative conditions.

By manipulating differentiation, they aim to create cells that can replace damaged tissue and restore function in patients.

Challenges and Future Directions

While our understanding of cell specialization has advanced, there are still mysteries to solve. One challenge is guiding cells to differentiate precisely without unwanted mutations or side effects.

Another is developing techniques to control differentiation reliably and at scale, which is crucial for therapeutic applications.

In the future, as we continue to decode the intricacies of cell specialization, we may see xenobots and other bioengineered life forms used in medicine, environmental cleanup, and beyond. Imagine xenobots delivering drugs directly to a tumor, or engineered cells repairing damaged organs at a cellular level.

Understanding the secrets of differentiation holds the key to such advancements, pushing the boundaries of both biology and technology.

In sum, cell specialization and differentiation are foundational to life and hold incredible potential for innovation.

By unlocking these processes, scientists can bridge the gap between nature and technology, creating tools that improve human health, the environment, and our understanding of life itself.

Chapter 4: Computational Design and Simulation

In the world of synthetic biology, the journey from concept to creation involves not just biological ingenuity but also computational finesse. Designing tiny, programmable organisms such as xenobots is no easy feat, and computational design and simulation play a critical role in making this visionary work a reality.

By exploring the intricate landscape of digital modeling and algorithmic precision, scientists are shaping cellular forms that nature itself never designed.

Through computer-aided design (CAD) and advanced simulations, xenobots can be carefully crafted and tested in virtual environments long before they're brought to life in a lab.

The Foundations of Computational Design

At its core, computational design for xenobots combines biological principles with engineering precision. It starts with the digital blueprinting of cells and tissues, a task that requires a deep understanding of how biological cells interact and form complex structures.

Each cell in a xenobot—whether it's muscle, skin, or a sensor cell—has its own functions and characteristics, much like individual components of a machine. Computational design translates these biological features into a digital format, enabling scientists to "see" how cells may come together to create a desired shape or function.

The first step in this design process is to select the specific cellular configurations that will give rise to the xenobot's unique form and abilities. Scientists use computer programs to model the possible arrangements of these cells.

A major consideration is how these cells will cooperate to produce movement, respond to environmental stimuli, and perform tasks such as carrying a payload or even repairing tissue damage.

While this may sound straightforward, it requires a sophisticated understanding of how cells behave at a microscopic scale, how they communicate, and how they respond to mechanical forces and chemical signals.

Virtual Testing Grounds: Simulation Environments

Once the initial design is mapped out, the next stage is simulation. Simulation acts as a virtual testing ground where scientists can refine their designs, assessing whether a particular cellular arrangement will behave as intended in a dynamic, real-world setting.

Think of it as a digital playground, where models of xenobots can "live," move, and interact with simulated environments. The software used for these simulations can account for various factors such as fluid dynamics (how the xenobot moves in water) and material properties (how the cells stretch, contract, and interact with each other).

To achieve realistic simulations, scientists use algorithms inspired by natural biological processes. For example, genetic algorithms are used to "evolve" xenobot designs by creating numerous design variations, selecting the best performers, and then recombining them to generate new designs.

This process mimics natural selection and evolution, enabling the program to "learn" which shapes and structures are optimal for specific tasks. The outcome is a design that is not only theoretically effective but also capable of adapting and functioning in real-life conditions.

Another key aspect of simulation is considering the constraints imposed by biology. Unlike conventional robots, xenobots are built from living cells, which have their own limits in terms of movement, durability, and environmental resilience.

The simulation environment must therefore include biological realism, meaning it considers how cells grow, divide, and respond to mechanical stress.

This is critical because a design that works in a computer program without accounting for these realities might fail when implemented in the lab.

Machine Learning and AI in Xenobot Design

Artificial intelligence (AI) is a powerful ally in the computational design of xenobots. Machine learning algorithms are employed to analyze vast datasets and uncover patterns that human designers may not readily see.

Through repeated testing and evaluation, AI algorithms can sift through thousands of xenobot configurations and suggest the most promising designs.

This use of AI is particularly valuable in optimizing xenobot functions, such as locomotion and task performance.

For example, by training a neural network on how different shapes and cellular arrangements impact a xenobot's ability to swim, scientists can quickly identify designs that offer superior movement efficiency.

The AI can even propose innovative solutions—such as novel cell arrangements or unique ways to organize muscle tissue—that would be difficult to imagine without computational assistance.

Moreover, AI-driven simulation allows for "self-optimization," where the system autonomously refines xenobot designs through trial and error, adapting to meet specific goals or overcome unforeseen challenges.

This capability pushes xenobot research into the realm of "self-designing" organisms, where the design software itself plays an active role in developing the best possible cellular configurations.

Feedback Loops: Closing the Gap Between Design and Reality

A central advantage of computational simulation is its ability to provide feedback that can inform and refine the design process. During simulations, scientists can monitor how well the xenobot prototype is performing in various tasks and use this data to adjust the design accordingly.

If, for example, a xenobot intended to move in a straight line is found veering off course due to an imbalance in its muscle cells, the design can be adjusted in the computer before being attempted in a lab.

This feedback loop between design and simulation allows for rapid prototyping, reducing the need for costly and time-intensive laboratory trials. Instead of spending months or years testing designs in a lab, scientists can run multiple virtual tests in a matter of hours, ultimately arriving at a near-perfect design before moving to the actual creation stage.

Applications of Computational Design and Simulation in Xenobot Research

The applications of this computational approach extend beyond simple experimentation. With the help of simulation, scientists can design xenobots with real-world applications, from environmental cleanup to targeted drug delivery.

For instance, computational simulations can test how effectively a xenobot design can pick up and transport microparticles in a fluid medium, potentially useful for cleaning up microplastics in water. Similarly, simulations can explore how xenobots might navigate through tissue environments, useful for applications in regenerative medicine.

In a broader sense, computational design and simulation are paving the way for a new era of bio-hybrid technology. The lessons learned from designing xenobots can be applied to other types of bio-robots and bio-machines.

This includes creating biological systems that are more integrated with human health technology, which could one day lead to advanced therapies for chronic diseases, non-invasive surgical options, and even personalized cellular robots for individual medical treatments.

A Marriage of Biology and Technology

Computational design and simulation serve as the foundation of xenobot research, bridging the gap between what we imagine in the digital world and what we can create in the physical world.

This fusion of biology and computer science is not only groundbreaking but also essential for developing functional, safe, and effective bio-robots.

By harnessing the power of computational design and simulation, scientists are building the future, one xenobot at a time, exploring the vast potential of programmable life in a way that was once purely science fiction.

How computer simulations aid in designing xenobots.

Computer simulations are a groundbreaking tool in the design and development of xenobots.

Xenobots, which are synthetic life forms created from living cells, have the potential to revolutionize fields like medicine, environmental science, and robotics.

These tiny, programmable biological machines have been engineered to perform specific tasks, such as moving in specific patterns, transporting tiny objects, or even self-repairing.

This remarkable capacity for precision control is possible largely due to advanced computer simulations that allow scientists to test, refine, and optimize designs before real-world experimentation begins.

1. Conceptualizing Xenobots: The First Step in Design

Computer simulations begin with conceptualization—where scientists and engineers visualize the characteristics and tasks a xenobot should perform. This involves setting parameters for the xenobot's environment, movements, and intended functions.

By establishing a virtual blueprint, scientists can test countless design possibilities before physically creating the xenobot. This saves both time and resources and enables precise tuning of various biological and mechanical aspects without needing to experiment directly on live cells.

For example, if scientists aim to create a xenobot capable of navigating a maze, they can simulate a maze environment on a computer and experiment with different shapes, cell placements, and movement patterns to determine the best design for achieving that goal. By modifying these variables in a virtual space, they can observe which designs are most effective and adapt accordingly.

2. The Use of Evolutionary Algorithms in Design Optimization

One powerful computational tool used in xenobot simulations is the evolutionary algorithm. Evolutionary algorithms mimic the process of natural selection by starting with a wide variety of random xenobot designs, then iteratively selecting and refining those that best achieve the desired outcomes.

Each cycle of refinement, called a "generation," takes the designs with the most promising qualities and combines them to create new "offspring" designs. Over successive generations, the algorithm yields xenobot configurations that are optimized for specific tasks.

For instance, if the goal is for a xenobot to efficiently move through a fluid, the algorithm will favor designs that exhibit faster, smoother movements in simulated fluid environments. Designs that perform poorly are discarded, while the most effective ones are kept, slightly altered, and re-tested in subsequent generations. This iterative process results in a finely tuned design that would be nearly impossible to arrive at by trial and error in physical experiments.

3. Simulating Cellular Interactions and Structure Formation

Xenobots are built using biological cells, typically from organisms like frogs, which makes understanding cellular behaviors essential. Computer simulations play a vital role in studying and predicting how cells will interact when arranged in different structures.

For example, if muscle cells are used, simulations help scientists predict how they will contract and generate movement. Similarly, skin cells can be tested virtually to see how they form stable outer layers that protect the inner structure.

Simulating cellular interactions provides insights into tissue properties and mechanical behavior, allowing researchers to test how different cell types work together.

If a xenobot requires a particular structure to accomplish a task, simulations allow scientists to determine whether that structure can be achieved with the cells available.

For instance, if a xenobot design needs flexibility, simulations can help determine whether muscle cells alone can provide the necessary movement or if a combination of cell types is required.

4. Testing and Refining Movements in Virtual Environments

Xenobots must be able to navigate environments to perform tasks, and simulating these environments allows researchers to test how xenobots interact with them.

For example, simulations can recreate water currents, gravity, or obstacles the xenobot might encounter. By observing how virtual xenobots move in these environments, scientists can refine their designs to ensure effective movement.

If a xenobot is meant to carry small particles, a computer simulation might test different movement patterns and structures to see which ones can support weight best and move with stability.

Virtual environments also allow for testing various control mechanisms, like "ciliary" movements, which mimic the way tiny hair-like structures on cells can propel them through fluid. Simulating these interactions helps ensure that, when xenobots are physically created, they have the movement capabilities needed for real-world applications.

5. Predicting Emergent Behaviors in Xenobot Assemblies

In many cases, xenobots are designed to work together in groups, where emergent behaviors can arise. Emergent behavior refers to the way systems display complex, collective behaviors that aren't explicitly programmed into individual parts.

For example, a group of xenobots might organize themselves into patterns or achieve coordinated movements that would be impossible for a single xenobot alone. Through simulations, researchers can predict and harness these behaviors to achieve specific outcomes.

When scientists simulate groups of xenobots, they can observe how individual actions contribute to group dynamics. If they want xenobots to self-organize into a line to push a larger object, they can test different interaction rules in a simulation to see which ones produce the desired effect. This provides valuable insights into how to engineer xenobots that can collaborate, adapting to new tasks as a team rather than as isolated units.

6. Validating Safety and Functionality Before Physical Testing

Since xenobots are made of living cells, ethical considerations and safety are paramount. Computer simulations provide a controlled environment where xenobot designs can be tested rigorously before any physical experimentation occurs.

Scientists can identify potential risks—such as unintended movements, instability, or unwanted interaction with biological tissues—well in advance.

This enables them to refine xenobot designs to minimize risks, creating safer models for real-world applications.

For example, if a xenobot is intended for a medical application, such as delivering medication to specific tissues in the human body, simulations allow researchers to ensure that the xenobot's movements are compatible with the body's fluids and tissues.

By testing different pathways and behaviors virtually, they can create a design that will not harm human tissues or introduce complications, ensuring that the xenobot performs safely and effectively in live environments.

7. Accelerating Innovation and Reducing Resource Consumption

One of the greatest advantages of computer simulations is the ability to rapidly iterate on xenobot designs without consuming physical resources.

Each iteration of a xenobot in simulation costs only computational power rather than biological materials, which are expensive and often limited.

This means that researchers can explore a vast range of possibilities—designing and testing hundreds or thousands of potential xenobot configurations—in a short time and with fewer resources than traditional biological experiments would require.

Additionally, simulations allow for real-time feedback and quick adjustments, making it easier to innovate and respond to new challenges.

If researchers discover that a xenobot design doesn't perform as expected, they can adjust it almost immediately in the virtual environment and re-test.

This accelerates the development process, making it possible to bring xenobot designs from concept to practical use more efficiently.

Computer Simulations as a Pillar of Xenobot Design

Computer simulations have revolutionized xenobot design, enabling scientists to create, test, and refine these tiny, programmable biological machines with unprecedented precision.

By providing a virtual testing ground, simulations allow for rapid iteration, rigorous safety checks, and highly optimized designs.

From conceptualizing tasks to refining movement, predicting group behaviors, and ensuring safety, simulations are indispensable to xenobot development, paving the way for a future where programmable living machines are commonplace in medicine, environmental protection, and beyond.

Optimizing Xenobot Shapes and Abilities Through Algorithms

In the realm of biotechnology and robotics, few innovations are as captivating and complex as the xenobot.

Made from living frog cells, these small, programmable biological machines are the result of a groundbreaking fusion of biology and computer science.

Although simple in structure, xenobots have demonstrated remarkable abilities, like self-repair, propulsion, and even collective behavior.

But how can such tiny, soft-bodied organisms be so intricately designed to perform specific tasks?

The answer lies in optimization algorithms, which enable researchers to shape xenobots and enhance their functionality for a wide array of applications.

The Need for Optimization: Why Shape Matters

For xenobots, shape is not merely an aesthetic consideration; it directly impacts their capabilities and efficiency. Different tasks require specific body shapes, sizes, and configurations. For instance, xenobots designed to swim require different shapes than those made to crawl.

A xenobot created to transport small objects may have appendages or cavities for holding cargo, while one designed for sensing and responding to environmental stimuli might need more surface area.

The precise design of these tiny organisms is crucial for maximizing their effectiveness in tasks, and even slight changes in their form can significantly influence their performance.

Traditional trial-and-error methods would take years to manually test each shape variation and determine its effectiveness. Instead, researchers employ optimization algorithms, which allow computers to sift through countless shape possibilities and simulate outcomes, identifying the best designs for specific purposes far more quickly.

How Optimization Algorithms Work: Evolving Shapes for Functionality

Optimization algorithms operate much like a form of artificial evolution, iterating on potential designs to improve xenobot shape and function.

Researchers begin by defining objectives based on the task xenobots are meant to accomplish, such as movement speed, durability, or load-carrying capacity.

Then, an algorithm generates hundreds or even thousands of possible shapes, each of which is evaluated based on these objectives. Successful shapes that meet or exceed certain criteria are kept and refined, while less effective ones are discarded.

This process repeats over numerous iterations, resulting in highly optimized shapes tailored to the xenobot's intended function.

One popular approach is the genetic algorithm (GA), which is inspired by natural selection. In a GA, initial xenobot shapes act as "parent" designs, which are then "mutated" or "crossbred" to produce new "offspring" shapes.

The best shapes are selected in each generation, gradually leading to an optimized design. By using such evolutionary principles, the algorithm can discover novel shapes that researchers may not have otherwise considered.

Simulating Reality: Virtual Environments and Fitness Testing

Once the algorithm has generated potential xenobot shapes, researchers use simulation software to test these shapes in virtual environments that mimic real-world conditions. Here, various forces, pressures, and obstacles are introduced to evaluate how each shape performs under stress. For example, simulations might test a xenobot's speed on different terrains or its ability to navigate complex mazes. This process, often referred to as "fitness testing," is vital for understanding how a particular shape will likely behave in actual biological or physical conditions.

Simulated environments also allow researchers to test shapes for tasks that would be difficult, expensive, or impractical to test in a physical lab. Instead of creating hundreds of xenobots only to see many fail, researchers can narrow down the pool to the best virtual candidates. In this way, simulations save resources while delivering scientifically rigorous insights into which shapes will succeed.

Machine Learning and Deep Neural Networks in Shape Optimization

Beyond traditional algorithms, more advanced techniques like machine learning (ML) and deep neural networks (DNNs) are increasingly being used to optimize xenobot shapes. These methods go a step further by recognizing patterns and predicting the success of certain shapes based on vast amounts of data. Unlike standard optimization algorithms that test each shape individually, ML models learn from previous outcomes to make smarter guesses in future iterations.

For instance, a DNN can process millions of simulations, learning to identify features that contribute to success in specific tasks. As the model learns, it can generate highly efficient designs faster and more accurately than traditional methods.

This predictive capacity makes DNNs ideal for complex optimizations, allowing researchers to explore previously unexplored designs that might have been dismissed due to their unconventional structure.

Multi-Objective Optimization: Balancing Trade-Offs for Versatile Xenobots

In many cases, xenobots must balance multiple objectives. For example, a xenobot might need to be fast, durable, and capable of self-repair. Multi-objective optimization algorithms are used to balance these competing needs, creating xenobot shapes that perform well across a range of criteria rather than excelling at just one.

Researchers employ techniques like Pareto optimization, which involves mapping different shapes along a spectrum of objectives to find those that offer the best trade-offs.

Through multi-objective optimization, xenobots can be designed for versatility, with shapes that are robust yet agile, fast but energy-efficient. This ability to balance trade-offs makes them highly adaptable to unpredictable environments, such as inside the human body for drug delivery or in polluted waters for cleanup.

Real-World Applications: Tailoring Xenobot Shapes for Specialized Tasks

The shape optimization process has already led to specialized xenobot designs with impressive real-world applications. In medicine, xenobots can be shaped to navigate through blood vessels, potentially carrying drugs to targeted areas or clearing blockages.

In environmental science, xenobots can be optimized for tasks like collecting microplastics or detecting contaminants in water.

The use of algorithms to tailor xenobot shapes is crucial for these applications, where precision and efficiency are essential. For example, xenobots designed for cleaning pollutants might have shapes that increase their surface area, allowing them to trap more particles. In contrast, xenobots intended for fast delivery through tight spaces may be slender and streamlined. Each application informs the shape and behavior of the xenobot, and optimization algorithms help researchers fine-tune them for these specific tasks.

The Future of Xenobot Design: Dynamic Shapes and Adaptive Capabilities

With continuous advancements in algorithmic techniques, xenobots may soon have the capacity for shape-shifting or adaptive behavior. Imagine xenobots that can alter their form to suit changing environmental conditions or transform their shape in response to obstacles. Future algorithms could integrate real-time feedback, enabling xenobots to adjust themselves autonomously for maximum effectiveness.

Dynamic shape optimization would open up entirely new possibilities for xenobots, including adaptability in unforeseen circumstances. Such xenobots could be used in disaster recovery, adjusting their shapes to navigate debris, or in medical applications where they might adapt to different bodily tissues or injuries.

A World Shaped by Algorithmic Precision

The marriage of biology and algorithmic science has propelled xenobot design from mere prototypes to specialized, purpose-driven tools. Through optimization algorithms, researchers can precisely control xenobot shapes, crafting living machines that push the boundaries of what is possible.

This technology represents more than just a scientific feat; it is a glimpse into a future where the line between the organic and the synthetic becomes a little more blurred, and where the potential for programmable life forms is as boundless as human imagination.

Chapter 5: Evolutionary Algorithms in Xenobot Design

In recent years, the field of robotics has taken a turn into truly organic territory with the advent of Xenobots—programmable, living organisms derived from frog cells. These tiny biological robots represent a breakthrough in understanding how life can be designed, reshaped, and directed toward novel functions.

Central to the development of Xenobots is the use of evolutionary algorithms, a set of computational techniques inspired by natural selection that allow researchers to design and optimize these living systems in ways that would be challenging or even impossible with traditional engineering.

What Are Evolutionary Algorithms?

Evolutionary algorithms are a class of computational methods that mimic the process of natural evolution. In nature, organisms evolve over generations, adapting to their environments and gradually optimizing traits that increase their chances of survival and reproduction.

Evolutionary algorithms apply this concept to problem-solving by generating a population of potential solutions and then iteratively refining these solutions through processes akin to mutation, selection, and reproduction.

In the context of Xenobot design, evolutionary algorithms enable researchers to create, test, and refine a variety of biological configurations. By simulating many generations of virtual Xenobot "offspring," researchers can discover designs that best fulfill a given function, such as movement, repair, or targeted drug delivery.

Rather than attempting to engineer specific solutions by hand, scientists rely on the power of these algorithms to explore the vast, complex landscape of possible designs more efficiently.

The Basics of Evolutionary Algorithms in Xenobot Design

In evolutionary algorithms used for Xenobot design, the process begins with a virtual environment where many potential configurations of cells are created. Each configuration, or "genotype," specifies the arrangement and properties of cells in the Xenobot.

The algorithm then assesses each genotype's performance in achieving a specific goal, such as moving in a particular direction, navigating an obstacle, or forming a specific shape.

1. Initialization: The first step is to initialize a population of random configurations. Each configuration will represent a unique arrangement of cells that could theoretically be built into a functional Xenobot.

2. Evaluation: In this step, the algorithm evaluates each configuration by simulating how the Xenobot would perform in the virtual environment. This might involve assessing how fast it moves, how efficiently it completes a task, or how stable its structure is over time.

3. Selection: After evaluating all configurations, the best-performing ones are selected to "survive" to the next generation. These are the designs that show the most promise in achieving the desired function.

4. Crossover and Mutation: The selected configurations are then used to create a new generation. Through crossover, aspects of two different configurations are combined to produce offspring, creating genetic diversity.

Mutation introduces small random changes to offspring, mimicking the biological mutation process, which can lead to novel and sometimes advantageous characteristics.

5. Iteration: This process is repeated over many generations, with each new generation ideally becoming progressively better at the given task. By the end of several hundred or even thousands of iterations, the evolutionary algorithm produces an optimized Xenobot design tailored to the desired function.

Why Use Evolutionary Algorithms for Xenobots?

Designing with cells rather than traditional materials presents unique challenges. Cells are highly dynamic, and their properties can change in response to their environment and interactions with other cells. This unpredictability makes it challenging to use conventional engineering methods.

Evolutionary algorithms, however, are well-suited to this complexity because they can explore countless possible configurations and "learn" which ones work best in the given conditions.

Evolutionary algorithms provide two main advantages in Xenobot design:

1. **Adaptability:** Unlike traditional robots, which must be carefully programmed for each task, Xenobots created through evolutionary algorithms can adapt to a range of scenarios.

For example, the algorithm might discover cell configurations that naturally self-organize to move in response to certain chemical signals or that can self-repair after being damaged.

2. **Discovery of Novel Forms and Functions:** Evolutionary algorithms can stumble upon unexpected solutions that a human designer might not consider. In some cases, the algorithm might develop entirely new types of movement or structural configurations that would be difficult to predict with conventional methods.

This openness to creative exploration is key for Xenobot research, as it allows scientists to uncover new ways cells can work together in dynamic, purposeful ways.

Evolutionary Algorithms in Action: Examples of Xenobot Functions

One of the most compelling aspects of using evolutionary algorithms is the range of functional Xenobots they can produce.

For example:

- **Movement:** One of the initial challenges in Xenobot design was creating a configuration that could reliably move across a surface. Evolutionary algorithms enabled researchers to test countless arrangements until they found one that allowed Xenobots to propel themselves forward through synchronized cell movements.

- **Collective Behavior:** Another application involved designing Xenobots that could work together to complete tasks. By setting a fitness function that rewards configurations capable of gathering small particles, researchers used evolutionary algorithms to develop Xenobots that collectively push objects to a central location.

- **Self-Healing:** Some Xenobots were designed with the ability to repair minor damage. By introducing evolutionary criteria that favored designs maintaining structural integrity after being punctured, scientists produced configurations that could spontaneously rearrange cells to close wounds, thereby prolonging the Xenobot's functionality.

Challenges and Future Directions

While evolutionary algorithms have been successful in generating functional Xenobot designs, several challenges remain. First, the computational power required for simulating and evaluating each generation is substantial, especially as designs become more complex.

Faster, more efficient simulations will likely be necessary as researchers strive to create more sophisticated Xenobots with finer control over behaviors and functions.

Additionally, transferring a successful virtual design to a real-world Xenobot is not always straightforward. Cells in a living organism may not behave exactly as they do in simulations, requiring additional fine-tuning.

This gap between virtual and physical performance highlights the need for advanced simulation techniques that accurately model cellular behavior.

The Promise of Evolutionary Algorithms in Synthetic Biology

The use of evolutionary algorithms for Xenobot design marks an exciting intersection between biology, computer science, and robotics. By enabling the discovery and refinement of bioengineered life forms with specific functions, evolutionary algorithms offer a glimpse into a future where synthetic biology and robotics merge seamlessly.

These algorithms hold promise not only for building Xenobots but also for exploring a range of biohybrid technologies that could transform medicine, environmental management, and beyond.

The design of Xenobots through evolutionary algorithms represents a leap forward in both technology and biology. These algorithms allow scientists to leverage the principles of natural selection to create innovative, functional biological machines. As the computational power and biological insights grow, so too will the capabilities of Xenobots, bringing us closer to a future where programmable life forms carry out tasks that were once thought impossible.

The journey of Xenobot design, fueled by evolutionary algorithms, is still in its early days, but it holds vast potential for revolutionizing our understanding and control of life at its most fundamental level.

Mimicking Natural Selection to Evolve Xenobot Designs

In the vast intersection of biology, robotics, and artificial intelligence lies a field teeming with potential: the study and development of "xenobots." These remarkable, tiny living machines are the product of biological cells programmed to act as miniature robots. Named after the African clawed frog Xenopus laevis from which their stem cells are harvested, xenobots open a fascinating chapter in science as they mimic aspects of living organisms.

But more intriguing is how scientists have begun to evolve these designs by mimicking a process that has governed life itself—natural selection. Let's explore how this evolutionary approach is unlocking the ability to develop efficient, adaptive xenobots with capabilities suited for specific purposes.

Understanding Xenobots and Natural Selection

Before diving into evolution in xenobots, it's essential to understand what xenobots are. Unlike traditional robots made of metal and electronic parts, xenobots are composed of living cells—specifically, stem cells.

These cells are guided to take on shapes and configurations that allow them to carry out various tasks, such as moving through fluid environments, transporting tiny objects, or even working together in groups.

The appeal of xenobots lies in their programmable nature, which allows scientists to design tiny "biobots" that can perform complex behaviors in controlled settings.

Natural selection, the mechanism first put forth by Charles Darwin, is the process by which species adapt over generations to become more suited to their environment. Traits that improve survival and reproduction chances are passed on, while less beneficial traits are phased out.

To evolve xenobots, scientists apply similar principles in a controlled laboratory setting, with powerful computer algorithms simulating the natural selection process. This approach allows them to explore countless designs quickly, choosing ones that offer the best performance in specific tasks.

Computational Evolution of Xenobot Designs

Creating a xenobot design through computational evolution involves simulating potential shapes, structures, and functions on a computer.

Researchers start with thousands of randomly generated xenobot designs, each a unique configuration of simulated stem cells that can act in coordination.

An algorithm evaluates each design's effectiveness based on specific criteria, such as speed, ability to transport objects, or endurance.

The process is iterative and guided by fitness selection, similar to natural evolution. The algorithm ranks each xenobot design by its "fitness" in the task it is meant to accomplish. Designs that perform better are "selected" to pass on their attributes to the next generation.

The best-performing designs are allowed to "breed" by combining and mutating their structural attributes, producing a new set of xenobot designs.

These new designs are then evaluated again, and the cycle repeats until the algorithm converges on optimal or near-optimal designs.

This evolutionary process enables scientists to discover designs that would be unlikely to emerge through human intuition alone. Given that xenobots are composed of living cells and can adapt to different tasks, the evolutionary approach allows for novel forms that are particularly well-suited to specific tasks, much like how natural organisms are adapted to their environments.

Mimicking Mutation and Selection Mechanisms

A fundamental aspect of natural selection is mutation, where random changes in DNA introduce new traits.

In computational evolution, mutation occurs when designs are altered slightly from one generation to the next, creating diversity in the design pool.

For xenobots, mutations might involve altering cell arrangements, changing proportions, or modifying movement mechanisms.

Another important mechanism is selection pressure, which drives evolution by favoring designs that meet desired criteria. In xenobot design, this pressure can be adjusted depending on the task.

For example, if the goal is to create a xenobot that moves rapidly, designs with faster, more efficient movement will receive a higher score. Over multiple generations, these selection pressures encourage the emergence of xenobot configurations that excel in specific functions.

Real-World Testing and Refinement

Once the computational process identifies promising designs, scientists translate these virtual blueprints into real xenobots. Stem cells from Xenopus laevis are harvested and guided to self-assemble into the specified configurations. Although virtual designs are promising, translating them into biological xenobots can be challenging.

Cells may behave differently in a physical environment than in a simulation. However, by iterating between computational simulations and real-world tests, scientists refine both the design and the biological assembly process to produce functional xenobots that closely match their simulated counterparts.

Applications of Evolved Xenobot Designs

Xenobots have an array of potential applications, from environmental cleanup to targeted drug delivery. For instance, one type of xenobot design might be tailored to detect and degrade microplastics in the ocean. Another design could carry drugs directly to infected tissues in the human body, reducing side effects and enhancing treatment precision.

One particularly promising application is in regenerative medicine. Xenobots could be engineered to help heal wounds by transporting cells to a damaged area, offering targeted support for recovery.

Since xenobots are biodegradable, they also present fewer ethical and environmental concerns than traditional robots.

The advantage of using evolutionary algorithms to design xenobots lies in the vast diversity of configurations they can create.

Designs that humans might not conceive, such as unusual shapes or unconventional movement patterns, can emerge naturally in an evolutionary process.

These novel configurations allow xenobots to perform tasks more efficiently than if they were designed by traditional engineering approaches.

The Future of Evolving Xenobot Designs

The evolution of xenobot designs is still in its early stages, but the field holds tremendous promise. As computer algorithms grow more powerful, and as we improve our understanding of cellular behaviors, scientists may unlock even more sophisticated design capabilities. In the future, we might see xenobots that adapt autonomously to their surroundings, responding to changes in their environment much like natural organisms.

Advancements in artificial intelligence and machine learning will likely play a significant role in this process. As algorithms become better at predicting and modeling cellular behaviors, xenobot designs will become even more refined, efficient, and specialized. This evolution could lead to xenobots that are capable of learning and adapting on their own, blurring the line between living organisms and machines even further.

Ethical Considerations in Xenobot Evolution

As xenobots evolve, ethical considerations are critical. What are the implications of creating self-replicating or autonomous living machines? Are there risks associated with deploying xenobots in natural ecosystems? Scientists are exploring these questions, striving to balance innovation with responsibility.

By mimicking natural selection to evolve xenobot designs, scientists are not just creating biological machines; they're unlocking new ways to understand and harness life itself.

The process of evolving xenobots demonstrates how principles that govern all life can be applied to create advanced, purpose-driven designs.

This remarkable blend of biology, AI, and robotics is paving the way for a future where living machines can solve some of our most complex challenges.

The journey has just begun, but each generation of evolved xenobots brings us closer to a new paradigm in science and technology.

Enhancing Xenobot Performance Through Iterative Processes

In the burgeoning field of bioengineering, few discoveries have stirred as much excitement as xenobots—tiny, programmable biological organisms that hold potential for a multitude of applications, from environmental cleanup to regenerative medicine. Composed of living cells from the African clawed frog (Xenopus laevis), xenobots are uniquely positioned at the intersection of biology and technology.

Researchers envision these micro-organisms as tools that could one day be harnessed for precise tasks, such as seeking out and removing harmful bacteria, delivering targeted drug therapies, or even aiding in tissue repair. But to realize these ambitious applications, it's essential to improve xenobot performance.

A promising approach to achieve this is through iterative processes, where scientists refine and optimize xenobot functions in successive steps. In this chapter, we delve into how iterative methods are enhancing xenobot design and functionality, breaking down complex processes into human-friendly explanations.

What Makes Iterative Processes Ideal for Xenobot Development?

At its core, an iterative process is one that relies on repetition to gradually improve a system. This approach is common in fields like engineering, artificial intelligence, and software development, where repeated cycles of design, testing, and refinement help optimize outcomes. The iterative process is well-suited to xenobot development for several reasons. Firstly, xenobots are composed of living cells, which can exhibit unpredictable behaviors and adapt in unexpected ways. By iteratively testing and refining xenobot designs, scientists can make gradual improvements and fine-tune how these organisms behave.

Moreover, the field of xenobot research is in its infancy, meaning that many initial designs have not yet been optimized. Iteration allows researchers to identify and address design flaws early on, minimizing wasted resources and maximizing the potential for innovative breakthroughs. This continuous refinement is crucial for evolving xenobots from experimental novelties to practical, robust solutions.

How Iterative Processes Work in Xenobot Engineering

1. Design Phase: The first step in enhancing xenobot performance through iteration is the design phase. Scientists use computer models and simulations to create initial blueprints for xenobots. These digital designs specify the types of cells that will be used, how they will be arranged, and the anticipated functions of the resulting xenobot.

Simulations are essential here, as they allow researchers to predict how different configurations might behave without the need for actual biological materials. In this way, the design phase acts as a preliminary filter, helping to eliminate ineffective models early on.

2. Fabrication Phase: Once a promising design has been identified, researchers proceed to the fabrication phase, where the xenobot is physically assembled using live cells.

Cells from the frog embryo are carefully selected and combined, often using micro-surgical tools or 3D bioprinting techniques to create the intended shape and structure.

For instance, heart muscle cells might be chosen for their contractile abilities, allowing the xenobot to exhibit movement, while skin cells offer structural integrity. This phase is delicate, as the cells need to be arranged with precision to achieve the desired functionality.

3. Testing and Observation Phase: After fabrication, xenobots undergo rigorous testing to assess their performance. Researchers observe how these organisms respond to different environmental stimuli and measure their ability to perform specific tasks.

For example, they may examine how well a xenobot can navigate through a maze, transport a small object, or aggregate with other xenobots. The data gathered during this testing phase is critical for identifying areas of improvement.

If a xenobot fails to perform a task as expected, researchers can pinpoint the limitations—whether they stem from structural flaws, cellular properties, or environmental incompatibilities.

4. Refinement and Reiteration: Based on test results, scientists refine the xenobot's design, adjusting cellular arrangements, material compositions, or structural layouts. They may, for example, add more contractile cells to improve mobility, alter cell placement to enhance stability, or modify the overall shape to facilitate navigation.

After making these adjustments, researchers fabricate and test the revised xenobot, repeating the cycle as necessary until performance improves.

Examples of Iterative Enhancements in Xenobot Functionality

The iterative approach has led to several advancements in xenobot capabilities:

- **Improved Mobility and Control:** Early xenobot models exhibited limited movement. Iterative adjustments, such as refining cell placement and using a higher proportion of contractile cells, have enabled xenobots to move more purposefully and with greater precision.

Some iterations have even incorporated neural cells, allowing for a degree of responsiveness to environmental cues, further expanding xenobot control.

- **Enhanced Longevity:** Initial xenobot designs had limited lifespans, as they were prone to cellular degradation over time. Through iterative experimentation, researchers have found ways to optimize the survival of these cells, allowing xenobots to function longer before breaking down. Adjusting the cell types or optimizing the external environment (such as pH and temperature) has extended their viability, making them more useful for prolonged tasks.

- **Swarm Intelligence:** One of the more ambitious goals in xenobot research is to create swarms—groups of xenobots that can work together toward a common goal. Through iterative processes, researchers have been able to improve xenobot aggregation and interaction.

By adjusting their size, shape, and signaling pathways, xenobots in later iterations can now coordinate better, showing promise for tasks like collective cleanup or assembly work.

The Role of Machine Learning in Iterative Improvement

Machine learning (ML) is also playing a transformative role in xenobot development by accelerating the iterative process. Algorithms can analyze the vast data sets generated from xenobot trials, identifying patterns and predicting the outcomes of potential design modifications. With ML-driven optimization, researchers can conduct thousands of simulated iterations rapidly, far outpacing the manual process.

For example, ML can suggest novel shapes or cell arrangements that may improve movement efficiency or stability—recommendations that scientists can then test and refine.

This collaboration between artificial intelligence and bioengineering has led to rapid strides in xenobot design. By harnessing the computational power of ML, scientists can explore a broader range of design possibilities, pushing the boundaries of what xenobots can achieve.

Challenges and Future Prospects

While iterative processes are invaluable in advancing xenobot performance, they also present challenges. Biological variability, for instance, means that not every xenobot will behave identically, even with identical designs. This inherent unpredictability can complicate testing and necessitate more iterations to ensure consistency.

Ethical considerations are another dimension, as xenobots are living organisms, albeit simple ones. As xenobots become increasingly sophisticated, questions about their status and treatment will likely arise, necessitating careful ethical scrutiny.

Despite these challenges, the iterative approach remains a cornerstone of xenobot research. The ability to make gradual, evidence-based improvements is enabling scientists to push the frontiers of this exciting technology, moving from theoretical potential to practical applications. With continued research and refinement, xenobots may one day become reliable tools for everything from medical therapies to environmental remediation, underscoring the power of iteration in the development of groundbreaking technologies.

Chapter 6: Xenobot Locomotion and Mobility

Imagine a microscopic creature that moves on its own, navigating fluid environments, exploring new terrain, and adapting its movement to changes in the environment. But this isn't a biological creature with DNA designed by evolution; instead, it's a xenobot—a programmable, organic robot made from living frog cells.

Xenobots, named after the African clawed frog Xenopus laevis, have captivated scientists because of their ability to accomplish autonomous movement using biological tissues without any metallic or synthetic components.

This chapter explores the mechanisms behind xenobot locomotion, the biology that powers their mobility, and the potential uses for this unique type of movement in medicine, environmental cleanup, and beyond.

1. Understanding Xenobot Locomotion

Xenobot movement is fundamentally different from what we see in typical robots, as it combines biological and robotic principles. Unlike traditional robots, xenobots aren't driven by wheels, gears, or an internal battery.

Their movement arises from two primary cell types: heart muscle cells (or cardiomyocytes) and skin cells, which work together to generate motion.

Heart cells in xenobots contract rhythmically. When clusters of these cells pulse in a coordinated fashion, they create a wave-like action that propels the xenobot forward.

The process is similar to how heartbeats pump blood, but here it produces movement. When these heart cells are strategically placed in the xenobot's structure, they serve as tiny "engines," powering the xenobot to move through fluid environments.

2. Xenobot Shape and Its Role in Movement

Xenobot mobility isn't solely based on the cell types used; shape plays a crucial role in determining how a xenobot moves. For example, a xenobot with a flat, disk-like shape may glide smoothly across a surface, while a xenobot with a more compact, spherical design can maneuver in a bobbing or crawling motion.

By altering a xenobot's shape, scientists can influence its speed, direction, and type of movement. This ability to program mobility through shape opens up possibilities for creating xenobots suited for specific tasks, like carrying tiny payloads or following predetermined paths in complex environments.

Computer simulations play a significant role in shaping xenobot designs, allowing scientists to test how different shapes interact with heart cell contractions and environmental factors. By experimenting with various shapes in a virtual environment, researchers can fine-tune the design before creating physical xenobots in the lab.

3. Steering and Control of Xenobots

While xenobots are autonomous and lack an external control mechanism, their movement can be subtly influenced by adjusting the placement and density of heart cells, as well as the xenobot's shape. In some designs, heart cells are concentrated in certain areas, creating a directional force that biases the xenobot's movement in a preferred direction.

Environmental factors also play a role in directing xenobot motion. For instance, changes in temperature, chemical gradients, or light can affect the behavior of the heart cells and, in turn, influence xenobot movement. Scientists are exploring how to use these environmental cues to create responsive xenobots that change direction or speed in reaction to their surroundings.

This could be useful in situations where xenobots need to reach specific areas in a dynamic environment, such as delivering a therapeutic payload to a particular part of the human body.

4. Challenges in Xenobot Mobility

Creating a self-propelled, self-navigating biological robot presents several challenges. One of the main difficulties is maintaining a consistent movement pattern, as heart cells can behave unpredictably outside the controlled environment of a living body. Moreover, heart cells in xenobots will eventually stop beating, which limits the lifespan and mobility of the xenobot.

Scientists are also working on ways to make xenobot movement more predictable. As heart cells age or lose energy, their contractions weaken, causing the xenobot to slow down or even stop moving. This natural decay presents an engineering problem for applications requiring longer-lasting or more reliable mobility. To address this, researchers are experimenting with various structural designs and cell types, hoping to create xenobots that are more resilient and adaptable.

5. Applications of Xenobot Mobility

The unique locomotion abilities of xenobots make them suitable for a wide range of potential applications, especially in environments that are challenging for traditional robots.

Here are a few key areas where xenobot mobility could be groundbreaking:

- **Environmental Cleanup:** Due to their tiny size and autonomous movement, xenobots can be used to explore polluted waterways, collect microscopic contaminants, or transport harmful particles to safe areas. Their organic makeup means they won't add to pollution and can safely decompose once their task is complete.

- **Medical Applications:** The autonomous nature of xenobots makes them ideal for tasks inside the human body. They could, for instance, be used to deliver drugs precisely to where they're needed, minimizing side effects. Researchers are exploring the possibility of xenobots programmed to navigate towards cancer cells or target bacterial infections, providing a new method for highly localized treatments.

- **Microassembly and Repair:** Xenobots' ability to move in controlled ways opens up possibilities for microassembly tasks. They could be used to carry out delicate operations, such as constructing micro-structures in laboratory environments. Similarly, in the future, xenobots might be programmed to perform cellular repair tasks, assisting in the regeneration of damaged tissues.

6. Future Directions in Xenobot Locomotion

The field of xenobot research is still in its early stages, but scientists are exploring numerous ways to enhance their mobility. One promising direction involves creating xenobots with more complex types of movement by integrating different types of cells, like nerve cells, that could allow for sensory feedback and greater control over movement.

As understanding grows, scientists hope to design xenobots that can learn from their environments, adapting their motion over time and responding to new challenges. This adaptive mobility could make xenobots useful for a wider variety of tasks, from complex medical procedures to the exploration of hazardous environments where other robots cannot operate.

Additionally, researchers are exploring ways to make xenobots "self-healing" so that they can recover from minor damage, much like a biological organism. If successful, self-healing xenobots could provide consistent, reliable performance even in difficult environments, making them a valuable tool in fields that require long-lasting and versatile mobile systems.

7. Ethical Considerations

The ability of xenobots to move autonomously brings up important ethical questions. Unlike machines, xenobots are made of living cells, which raises concerns about their potential to interact with biological ecosystems. If xenobots were used widely, ensuring they do not disrupt natural environments or cause unintended consequences is essential. Researchers are taking these considerations seriously, developing protocols to ensure that xenobots are safe, biodegradable, and environmentally friendly.

The science of xenobot locomotion is a fascinating convergence of biology, robotics, and computer science, offering exciting possibilities for future applications. By understanding and controlling the way xenobots move, scientists are opening new doors for how tiny, programmable organic robots could perform tasks that were previously unimaginable. As this field advances, the potential to harness xenobot mobility promises to revolutionize medicine, environmental science, and beyond, changing the way we approach some of humanity's most pressing challenges.

Various Methods of Xenobot Movement: Crawling, Swimming, and More

In the evolving world of synthetic biology, the ability to make xenobots—tiny, programmable living robots—move in specific ways is one of their most remarkable features. Unlike traditional robots, which are generally crafted from metal or plastic and moved by mechanical motors, xenobots are living structures made from the skin and heart cells of the African clawed frog (Xenopus laevis). These biological "bots" have their own intrinsic ways of moving, powered by natural cellular functions rather than wires and motors. The movement methods of xenobots can be categorized into distinct modes, such as crawling and swimming, with each type of motion offering unique advantages and challenges for potential applications.

Let's dive into the fascinating world of xenobot movement and see how these tiny living robots accomplish complex feats in motion.

1. Crawling: The Power of Cellular Squeezes

Crawling is one of the primary ways that xenobots navigate their surroundings, particularly when they are on a surface or substrate. This type of movement is achieved largely due to the interaction between the skin and heart muscle cells that make up these xenobots. The skin cells provide a structural framework, while the heart muscle cells act as mini powerhouses, contracting rhythmically. When these cells contract and relax in a coordinated pattern, the xenobot's body shifts forward in a crawling motion.

The crawling motion is primarily driven by localized contractions of heart muscle cells, which produce a wave-like pattern that can propel the xenobot forward in short bursts. This method is not particularly fast, but it is efficient for close contact movement on solid surfaces. Xenobots are capable of sensing physical cues from their environment, such as the texture of a surface, which can influence how they crawl. For instance, a smooth surface may enable a faster, more gliding crawl, whereas a rough surface could require the xenobot to exert more force, resulting in a slower but deliberate forward motion.

This crawling movement has potential applications for environments that require careful navigation, such as inside blood vessels or through dense tissue in a medical setting. Here, a xenobot's crawling ability would allow it to gently but persistently move through confined spaces, potentially carrying medicines or performing microscopic surgeries without the need for a rigid body or complex propulsion systems.

2. Swimming: Propelling with Cellular Pulses

Swimming is another fascinating form of xenobot movement, typically used when the xenobots are placed in a fluid medium, such as water. Unlike crawling, swimming requires a more dynamic and coordinated pattern of movement.

To swim, xenobots rely on the pulsating contractions of their heart cells, which can generate small but effective waves in the surrounding fluid, propelling them forward.

The swimming motion relies on the xenobot's overall shape and the distribution of contractile cells. For example, a rounded xenobot with a strategically positioned band of muscle cells can produce tiny ripples in the water, moving itself forward.

In some designs, researchers have shaped xenobots with tiny cilia-like structures or appendages to help direct these fluid motions, making their swimming more efficient.

This cilia-like movement allows xenobots to swim in various directions, even turning and adjusting course.

Swimming offers xenobots a greater degree of mobility and flexibility compared to crawling, making them suitable for tasks that require rapid relocation or navigation through liquid environments. In future applications, swimming xenobots could navigate the bloodstream, explore water systems, or even monitor aquatic ecosystems.

Their efficient, self-powered movement means that they could perform tasks without needing external power sources or complex navigation systems, making them ideal for environments where traditional machinery would struggle.

3. Rolling and Rotating: A Unique Way to Tackle Terrain

A less common but innovative form of xenobot movement is rolling or rotating. By manipulating the body design and cell placement, researchers have managed to engineer xenobots that can roll across surfaces. Rolling is achieved by shaping the xenobot in a spherical or cylindrical form, where contractions of the cells on one side of the body push it over itself, creating a rolling motion.

This movement is efficient for getting over certain obstacles or covering long distances with minimal energy use.

Rolling xenobots can be thought of as self-rotating spheres or barrels, leveraging gravity and controlled contractions to maintain their trajectory. This form of movement could be valuable in uneven terrain, where other movement forms, like crawling, might be slower or less stable.

Additionally, rolling could be useful in microfluidic systems, where the ability to navigate twists and turns quickly could provide advantages in medical diagnostics or environmental monitoring.

4. Pivoting and Turning: Navigating with Precision

Precision in movement is crucial for many potential applications of xenobots, and pivoting or turning movements are essential for this. Pivoting allows xenobots to make sharp directional changes by concentrating contraction forces on one side of their body, causing a controlled shift in orientation.

By incorporating asymmetrical designs, researchers enable xenobots to pivot and turn when needed. This movement is particularly important in confined spaces, where simply crawling or swimming forward isn't enough.

The xenobot's ability to reorient itself without needing to rely on external sensors is a testament to the flexibility that cellular contractions can provide.

Turning and pivoting open doors to complex navigation tasks, allowing xenobots to explore environments with obstacles, complex mazes, or targeted zones where they need to perform specific tasks.

For instance, in a medical application, xenobots that can pivot might be able to move directly to a damaged tissue area for precise drug delivery.

5. Aggregated Motion: Collaborative Swarming

In addition to individual movements like crawling or swimming, xenobots can also move in groups, demonstrating a kind of swarm-like behavior.

This aggregated motion occurs when multiple xenobots are released together in a fluid or surface, and they naturally form collective patterns.

Aggregated motion is largely due to the interactions between individual xenobots and their environment, where local signals or physical proximity encourage them to move together as a cluster.

Swarming has a range of potential applications, particularly for tasks that require coordinated action over a larger area, such as cleaning up toxic spills, collecting cellular debris, or acting as a bio-detector swarm.

The aggregated movement not only enables them to cover a broader area but also to function as a self-organizing system, capable of responding to environmental cues collectively without needing a central command.

A Glimpse into the Future of Xenobot Movement

The movement methods of xenobots represent more than simple biological curiosity; they are the first step toward creating programmable, adaptable biological systems with practical applications. Whether they're crawling across a surface, swimming through a liquid, rolling over terrain, pivoting with precision, or swarming in concert, xenobots offer a new vision of what biological machines can achieve.

The field of xenobot movement is still young, but as researchers continue to experiment with design and cell programming, we are likely to see even more complex and specialized forms of movement emerge.

By harnessing the natural behaviors of cells and directing them with purpose, scientists are charting a path toward biologically-based robotics with capabilities far beyond those of metal-and-motor machines. In the world of xenobots, movement is not just a means to an end—it's a frontier of scientific discovery, pushing the boundaries of what we understand about life, biology, and the future of robotics.

Challenges in Creating Efficient and Adaptable Locomotion Systems in Xenobots

The quest to create efficient and adaptable locomotion systems in xenobots is a fascinating intersection of biology, robotics, and artificial intelligence.

Unlike traditional machines, which rely on rigid and predictable movements, xenobots—living micro-organisms made from the cells of African clawed frogs (Xenopus laevis)—possess a more organic and flexible structure.

This flexibility enables them to perform a range of tasks but also introduces a suite of challenges in controlling and optimizing their movement.

Let's explore the major hurdles researchers face in designing xenobot locomotion systems that are both energy-efficient and adaptable to various environments.

1. Biological Constraints and Structural Limitations

At the core of a xenobot's structure are living cells, specifically frog skin and heart cells, which limit design possibilities. Skin cells provide a protective layer, while heart cells contract and expand, acting like tiny engines driving movement.

However, unlike mechanical motors, heart cells' contractions are not directly programmable. Researchers can encourage certain patterns, but they cannot yet fully control each individual cell's behavior with the same precision as a robotic actuator.

This biological variability makes it difficult to produce predictable and controlled locomotion patterns.

Furthermore, the fragility and short lifespan of living cells introduce limitations on long-term usage. Unlike metal or synthetic materials, cells are subject to degradation and may lose functionality over time. This challenge necessitates a careful balance between maintaining xenobot longevity and optimizing them for complex tasks, like navigation through varied terrain.

2. Achieving Energy Efficiency in Movement

Energy efficiency is a major challenge, as xenobots cannot carry or generate energy the way traditional robots might with batteries or fuel cells. Instead, they rely on the inherent energy within living cells, which is finite and diminishes over time.

Researchers are investigating how to make the best use of this cellular energy, aiming to design xenobots that can travel as far and as fast as possible without needing external energy sources.

One potential solution involves fine-tuning the contractile rhythm of heart cells, allowing for smoother and more efficient movement patterns. By manipulating cellular signals, researchers hope to synchronize these contractions, minimizing unnecessary energy expenditure.

However, achieving such fine-tuned coordination is challenging, as heart cells are notoriously difficult to control individually, and they often act in clusters that respond differently depending on their environment. The energy efficiency of xenobots, then, depends on optimizing these cellular contractions and synchronizing movement across many cells, a feat that still eludes current technology.

3. Adapting to Varied Environments

Xenobots are designed to operate in diverse environments, from liquid solutions to semi-solid gels. This versatility is both a strength and a complication, as each environment requires different forms of movement.

For example, in fluid environments, xenobots might need to swim by using synchronized cell contractions, whereas on semi-solid surfaces, they may require a different type of crawling or undulating motion. Adapting to these varied environments demands a level of environmental sensing and responsive movement that current xenobot designs are only beginning to achieve.

To address this, researchers are exploring adaptive cell programming, aiming to trigger specific behaviors in response to environmental cues. For example, a xenobot could sense changes in pH or temperature and adapt its movement accordingly. Although promising, this approach introduces another layer of complexity, as it requires xenobots to possess some degree of sensory capability—a characteristic that biological cells do not naturally have in a way useful for movement adaptation.

4. Control and Navigation Complexity

Traditional robots are equipped with sensors, processors, and communication systems that enable navigation and control. Xenobots, by contrast, lack any such circuitry or sensory apparatus. Without conventional control mechanisms, researchers must find innovative ways to guide xenobots in desired directions. Most current approaches rely on predefined shapes and cell arrangements to encourage specific movement patterns, but these configurations are limited in scope and cannot dynamically adapt to real-time obstacles or changing paths.

One approach to enhance control is through "morphological intelligence," a concept that uses a xenobot's shape to dictate its movement. By creating specific body shapes—such as C-shaped or spiral forms—scientists can guide xenobots to move in more predictable ways. Although morphological intelligence shows potential, it lacks the flexibility to respond to changing environments. Researchers are also experimenting with chemical cues and light-based guidance, although these methods are still in their infancy and add another layer of complexity to an already delicate system.

5. Designing for Resilience and Self-Healing

Living systems have a remarkable capacity for self-repair, a trait that researchers hope to leverage in xenobot design. However, fostering resilience and self-healing in a way that supports locomotion is an open challenge. When a xenobot sustains damage—such as a tear or cell death—it may lose its ability to move effectively or become completely immobilized. To counteract this, scientists are experimenting with regenerative cells that can heal small injuries, potentially extending the xenobot's lifespan and movement capacity.

Incorporating self-healing properties is no easy task, as it requires cells to respond to damage in ways that support continued movement. Some experimental designs use skin cells that can seal minor wounds, although this capability remains limited and unpredictable. As research advances, self-healing features could make xenobots more durable and capable of traversing rougher terrain, but achieving reliable self-repair for movement purposes remains a future goal.

6. Integrating AI for Autonomous Decision-Making

For xenobots to be effective in real-world applications, they may eventually need some level of autonomy and decision-making capacity. AI offers promising solutions, particularly in simulations where scientists can predict and optimize movement patterns. However, the leap from digital simulations to real-life xenobot locomotion is substantial. Researchers face the challenge of translating AI-derived movement patterns into the biological behavior of living cells, which are inherently variable and subject to external influences.

In simulated environments, AI algorithms can optimize xenobot shapes and predict likely movement paths. These simulations help researchers test multiple design iterations quickly, enabling a more efficient experimentation process.

However, transferring AI-optimized designs from virtual simulations to physical xenobots still faces biological constraints, as living cells do not behave as predictably as computer-modeled particles.

Current AI applications are therefore limited primarily to the design phase, although future breakthroughs could enable real-time AI adjustments to xenobot movement.

7. Ethical and Practical Implications

The development of adaptable and efficient locomotion in xenobots raises ethical questions as well. As these tiny organisms begin to show more complex behaviors, researchers and ethicists are debating how best to oversee and regulate xenobot research.

Issues around containment, environmental impact, and the possibility of unintended consequences must be considered, especially as xenobots become more autonomous.

Creating efficient and adaptable locomotion systems in xenobots remains a frontier filled with both promise and complexity. Each challenge—biological constraints, energy efficiency, environmental adaptation, control, resilience, and AI integration—presents unique obstacles.

Yet, with every advance, researchers move closer to realizing the full potential of xenobots in applications ranging from targeted drug delivery to environmental cleanup.

The path is demanding, but the rewards could transform our understanding of biology and robotics in profound ways, opening new horizons in the field of bioengineering.

Chapter 7: Sensing and Perception in Xenobots

In the ever-evolving field of biotechnology, xenobots represent an innovative leap into designing life forms capable of responding to their environment autonomously. Created from frog cells, these tiny, programmable organisms have fascinated scientists and the public alike due to their potential applications across various fields, from medicine to environmental cleanup.

One of the most intriguing aspects of xenobots is their ability to sense and perceive elements in their surroundings, enabling them to navigate and respond to different conditions in real-time.

Here, we dive into the sensory mechanisms that allow xenobots to perceive their world and discuss the scientific methods that make this perception possible.

Basic Sensing Abilities

At their core, xenobots are designed from skin and heart cells taken from the African clawed frog (Xenopus laevis), giving them the name "xenobots." While these cells do not inherently possess complex sensory abilities, scientists have cleverly utilized their natural properties to create rudimentary sensing capabilities.

Skin cells, for example, form the outer layer of the xenobot and act as a barrier, but they are also capable of responding to external physical stimuli.

Meanwhile, heart cells provide a source of movement. When grouped together in the right formation, these cells enable xenobots to move in response to certain environmental cues, functioning in a way that could be described as "sensing" even though they lack conventional sensory organs.

Mechanisms of Perception

Xenobots do not have brains or nerves, which limits their sensory perception as compared to organisms that process sensory information through nervous systems. Instead, they rely on a more distributed form of intelligence known as "embodied intelligence." This concept suggests that the xenobot's structure and cellular organization allow it to process environmental information directly through its cells and respond without requiring a central processing unit.

This form of intelligence allows xenobots to navigate their surroundings based on sensory input from environmental factors such as chemical gradients, physical obstacles, and, in some cases, light. For example, a xenobot might detect a chemical signal in the water and respond by moving towards or away from the source.

This reaction is not driven by conscious thought or complex processing but rather by direct cellular interaction with the environment, causing the cells to change behavior in response to specific cues.

Chemical Sensing in Xenobots

Chemical sensing is one of the most promising sensory applications in xenobots. By modifying the cellular makeup of xenobots, researchers can design them to detect and respond to certain chemical cues. This is achieved by embedding receptors within the cell membrane that react to specific molecules, such as toxins or pollutants. When a xenobot encounters a target chemical, the receptor binds to it, triggering a response in the cell that may result in movement toward or away from the substance.

This chemical sensing capability has significant potential in environmental science. For example, xenobots could be deployed in polluted water sources to detect harmful compounds, identifying contamination zones by moving toward them or congregating in specific areas.

In this way, xenobots could serve as a real-time, self-organizing monitoring system for ecological health.

Physical Sensing Through Shape and Motion

Xenobots can also respond to physical stimuli in their environment. The combination of heart cells and skin cells enables them to move in a coordinated manner, propelling themselves through their environment. Scientists have designed xenobots to have different shapes, which, in turn, affects their movement patterns. By altering their physical form, researchers can influence how a xenobot navigates and interacts with its surroundings. For example, xenobots with a spherical shape can roll, while those with a more elongated shape can "crawl" along surfaces.

This responsiveness to physical stimuli allows xenobots to maneuver around obstacles in their path. Although they lack a sophisticated sense of "touch" as we understand it, the cells respond to direct physical contact by changing direction or speed, allowing the xenobot to adapt its movement accordingly. In some experiments, xenobots have shown an ability to reconfigure themselves to move around obstacles, showcasing a primitive form of adaptive behavior.

Sensory Integration and Pattern Recognition

A xenobot's ability to interpret and act on environmental cues is possible due to pattern recognition at the cellular level. Although xenobots lack a nervous system, their cells can interact with one another in a way that enables them to function as a collective unit.

This cooperative behavior is reminiscent of the way simple multicellular organisms respond to external stimuli by relying on cellular communication.

For instance, in the presence of certain chemicals, individual cells in a xenobot can communicate through chemical signals or direct contact, altering the behavior of neighboring cells. This results in the xenobot "steering" itself toward or away from stimuli.

Pattern recognition in xenobots is far more rudimentary than in higher organisms, but it opens up possibilities for more complex responses if researchers are able to further develop the signaling pathways between cells.

Potential for Artificial Intelligence Integration

One of the most exciting future prospects for xenobot technology is the potential integration of artificial intelligence (AI) to enhance their sensory and perceptual abilities. AI could enable xenobots to better "learn" from their environment and make decisions that improve their efficiency in tasks like environmental monitoring or targeted drug delivery.

AI algorithms could be embedded in xenobot designs to respond in more sophisticated ways to multiple stimuli simultaneously.

Through AI, xenobots could, for example, develop the capability to detect a broader range of chemical signatures and respond differently to each type.

Over time, AI-enhanced xenobots might "learn" optimal paths for detecting specific toxins in water or identifying cellular damage within the human body for medical purposes.

The fusion of AI with biological systems in xenobots could allow them to carry out tasks that require more nuanced perception, adaptive behavior, and even basic decision-making.

Ethical and Practical Considerations

As we explore the sensory capabilities of xenobots, it's essential to address the ethical implications of creating and deploying bioengineered organisms with sensing and perceiving abilities.

Although xenobots are far from conscious beings, their ability to interact autonomously with the environment raises questions about control, environmental impact, and unintended consequences.

Scientists must consider containment strategies to ensure xenobots do not interfere negatively with natural ecosystems.

Additionally, developing sensing xenobots requires transparent guidelines to prevent misuse in scenarios such as surveillance or warfare.

The Future of Xenobot Perception

Xenobots' capacity for sensing and perception, though limited by the absence of a central nervous system, represents a fascinating frontier in biotechnology. By understanding and enhancing these rudimentary sensory abilities, scientists can open new possibilities for using xenobots in applications that require a biological interface with the environment.

Whether for environmental monitoring, medical applications, or even industrial tasks, the evolving perceptual abilities of xenobots will likely continue to redefine what bioengineered life forms can achieve in the years to come.

Incorporating Sensory Capabilities into Xenobots

A New Frontier in Robotics and Biology

As biological robots crafted from living cells, xenobots are a fascinating example of how far we've come in merging biology and technology. By combining frog-derived stem cells with the design power of artificial intelligence, scientists have created entities that can move, work together, and even heal themselves. But in order to make xenobots more versatile and intelligent, researchers are focusing on a critical advancement: the incorporation of sensory capabilities. Imagine tiny biological machines that can not only act in specific ways but also react to environmental cues, communicate, and respond to stimuli in real-time.

What Are Sensory Capabilities?

Sensory capabilities allow organisms to perceive and react to their surroundings. In humans and animals, senses include touch, sight, hearing, taste, and smell. Cells, on a much smaller scale, also possess sensory abilities. They can respond to chemical signals, detect physical contact, and even sense light. By incorporating similar sensory features into xenobots, scientists aim to create "smart" biobots that can understand and adapt to their environment.

For xenobots, sensory capabilities could mean the ability to detect chemicals, sense physical barriers, or even respond to light.

These capabilities could allow xenobots to navigate complex environments, locate specific targets, or avoid harmful areas, greatly enhancing their range of applications.

Why Are Sensory Capabilities Important for Xenobots?

Currently, xenobots are limited by their simplicity. While they can move and cluster together, they lack the ability to adjust their behavior based on environmental cues.

Sensory capabilities would allow xenobots to operate autonomously in more complex environments.

For instance, xenobots could be used to clean up polluted water by detecting and responding to specific chemical contaminants.

In medicine, xenobots could navigate the human body more effectively if they could sense and respond to physiological signals, allowing them to target infected or cancerous cells precisely.

This type of responsiveness is particularly important for tasks that require a high degree of adaptability.

Imagine a swarm of xenobots deployed to clear microplastics in the ocean; sensory capabilities could help them recognize and focus on microplastics while avoiding harmless debris.

Sensory-enabled xenobots could also avoid obstacles, navigate narrow pathways, or work together more effectively by sensing each other's presence, increasing their potential in both environmental and medical fields.

How Are Sensory Capabilities Integrated into Xenobots?

Incorporating sensory abilities into xenobots involves a multi-disciplinary approach that combines biology, engineering, and artificial intelligence.

Here's a breakdown of how sensory capabilities can be added to xenobots:

1. Cell Selection and Engineering:

Since xenobots are built using frog (Xenopus laevis) stem cells, researchers can select cells that naturally possess sensory abilities. For example, certain cells are responsive to light or can sense chemical gradients. By choosing these cells and incorporating them into the xenobot structure, researchers can create xenobots with inherent sensory abilities.

However, engineering these cells to enhance or customize their sensory functions is often required to achieve specific goals, such as detecting particular chemicals or responding to physical touch.

2. Genetic Modification:

Genetic engineering provides another way to add sensory features. By modifying the DNA of the cells used to build xenobots, scientists can insert genes that enable them to produce proteins linked to sensory functions. For instance, adding genes that code for opsins—light-sensitive proteins—would allow xenobots to respond to light. Similarly, genetic modifications could enable xenobots to react to particular chemical compounds, making them capable of tasks such as targeted drug delivery or pollutant detection.

3. AI-Driven Design and Training:

Artificial intelligence is crucial in designing xenobots with sensory capabilities. AI algorithms simulate various design structures and test how these designs might perform in specific environments. AI can also predict which sensory capabilities would be most effective for a given task and recommend modifications to enhance their effectiveness.

Machine learning models, for instance, can train xenobots to associate specific sensory inputs with actions, creating simple but effective reflexes within the xenobot design.

4. Neuromorphic Components:

Adding neuromorphic—or brain-like—elements to xenobots may also support sensory capabilities. These components can process sensory information in a way that resembles neural networks in biological organisms.

Neuromorphic elements might allow xenobots to "decide" to change direction when encountering a barrier or to cluster around a specific chemical signal. While rudimentary at this stage, these elements could help xenobots become more autonomous and responsive.

5. Biohybrid Systems:

Biohybrid systems combine living cells with non-living components, such as synthetic sensors. For example, scientists can attach small, synthetic sensors to xenobots to detect environmental changes. These sensors might emit electrical or chemical signals that the xenobot cells can interpret, allowing for a feedback loop. This approach allows for more complex sensory functions without the need for extensive genetic modification.

Challenges in Developing Sensory Xenobots

Incorporating sensory capabilities into xenobots presents several challenges. Biological materials are naturally variable, and it's difficult to create a consistent sensory response across all xenobots in a swarm.

Moreover, controlling how these sensors interact and ensuring that xenobots interpret sensory information accurately are major technical hurdles.

Another issue is energy. Xenobots, unlike traditional robots, lack a continuous power source. Any sensory function added must operate within the energy constraints of a cell-based system.

Scientists are exploring ways to make xenobots self-sustaining, possibly by harnessing cellular metabolism, but this remains a challenge.

Ethics also come into play. Xenobots with sensory and reactive capabilities are stepping toward a form of synthetic biology that raises questions about autonomy, the boundaries between living and non-living entities, and the potential risks if xenobots were to malfunction in sensitive environments.

Potential Applications of Sensory Xenobots

With sensory capabilities, xenobots could achieve impressive feats across multiple fields:

1. Environmental Monitoring and Cleanup:

Sensory-enabled xenobots could detect pollutants in water sources and work to neutralize them. They might also respond to environmental cues, such as temperature or pH levels, and adjust their behavior to optimize clean-up efforts.

2. Medicine and Surgery:

In medical applications, sensory xenobots could navigate the body to target specific cells or tissues. For example, they could identify and eliminate cancerous cells or release drugs at precise locations.

3. Search and Rescue Operations:
Sensory xenobots could navigate disaster sites, detecting chemicals that indicate human presence or locating areas too small for traditional robots. They could enter areas of debris and respond to signals that guide them to survivors.

The Future of Sensory Xenobots
The integration of sensory capabilities into xenobots is a promising and evolving field. As scientists make progress, the dream of creating fully autonomous, environmentally responsive biological machines is becoming a reality.

These advancements may one day enable xenobots to perform critical tasks that improve health, protect the environment, and open new possibilities for robotics and biology.

As we venture deeper into the Xenobot Chronicles, we'll witness how adding sensory abilities transforms xenobots from simple movers to sophisticated, responsive agents in the world of synthetic biology.

Using Sensors to Navigate and Interact with Their Environment

In the world of robotics, sensing and navigating within an environment are crucial for performing complex tasks. Traditional robots use electronic sensors—like cameras, infrared, and ultrasonic sensors—to map out their surroundings. Xenobots, however, have taken a novel approach: these biological machines, crafted from living cells, leverage intrinsic cellular behaviors and responsiveness to navigate and interact with their environment in ways that blur the line between biology and technology. Imagine a living robot sensing its environment without traditional electronics, relying instead on the cells' natural, evolved abilities.

Cellular Basis for Sensory Detection

Xenobots are not built from wires and circuit boards; they are made from biological tissues, often using frog embryonic cells—particularly skin and cardiac cells. Skin cells, known as ectodermal cells, are responsible for forming protective barriers and have developed mechanisms to detect environmental stimuli. Cardiac cells, on the other hand, are specialized muscle cells that beat rhythmically.

This natural beating can be harnessed to propel the xenobots forward, making movement a product of cellular function rather than motor-driven mechanics. Both cell types contribute to the xenobot's ability to sense and respond to environmental changes.

One of the fascinating features of living cells is their responsiveness to various stimuli. Cells naturally possess receptors that can detect chemical, electrical, and mechanical signals. By programming the formation and arrangement of specific cell types, scientists can create xenobots with an elementary "awareness" of their surroundings.

These robots respond to their environment in a manner similar to how organisms respond to food, light, or pressure.

Mechanisms of Sensing: Chemical, Mechanical, and Electrical Cues

Living cells communicate through chemical signals, a process known as chemotaxis. In xenobots, chemotaxis allows them to "smell" or detect chemical gradients in their environment.

By tuning this mechanism, scientists can direct xenobots to move toward or away from specific chemical sources.

For example, xenobots have been shown to migrate toward certain chemicals, which could be applied in real-world scenarios like targeting pollutants or areas of high toxicity for environmental cleanup.

This natural ability to detect and respond to chemical signals is reminiscent of the way bacteria and immune cells navigate within living organisms.

Mechanical cues also play a significant role. The flexibility of biological tissues allows xenobots to feel the contours of their environment.

When encountering obstacles, the xenobot's soft body can deform, allowing it to fit into crevices or avoid harmful impacts.

This tactile sensing, though basic, allows xenobots to physically adapt to varied terrains without complex algorithms.

Their softness gives them an edge in maneuvering through environments where rigid robots might get stuck or damaged.

Moreover, cells are sensitive to electrical signals. Bioelectricity—a natural phenomenon in all living cells—can be harnessed within xenobots to control their movement or direct them towards specific stimuli.

Researchers are exploring how altering bioelectric fields could prompt xenobots to behave differently, potentially allowing for more precise navigational control and task-specific adaptations.

Interactive Behavior: Communication and Collaboration

Xenobots can not only detect changes in their environment but also interact with it in ways that mimic simple forms of communication. When multiple xenobots are placed together, they can exhibit swarm behavior, working collectively without centralized control.

By following local cues—like gradients of chemicals or signals emitted by nearby xenobots—they can synchronize their actions and organize themselves into larger formations.

This group behavior, akin to swarming seen in certain animals, is a form of indirect communication through shared environmental signals.

This coordinated behavior holds potential for collective tasks, such as assembling materials, targeting large areas for environmental remediation, or even working within the human body to break down cellular debris.

Unlike traditional robots that require pre-programmed instructions for collaboration, xenobots naturally "cooperate" based on environmental cues, making them uniquely suited for unpredictable or dynamic tasks.

Navigating Complex Terrains

Xenobots possess an intrinsic advantage when it comes to navigating challenging terrains. Their soft, flexible structure means they can move through irregular environments, like rocky terrain or narrow passageways.

Unlike rigid robots, xenobots can "flow" around obstacles, utilizing their flexible bodies to adapt shape in response to the physical features of their surroundings.

This adaptability is crucial for applications in hard-to-reach or hazardous areas where traditional robots might face difficulty.

For instance, xenobots could traverse complex areas to perform delicate cleaning or maintenance tasks, especially where sensitivity to surroundings and adaptability are required.

Sensing for Health and Environmental Applications

One potential application for xenobots' sensory capabilities is in health monitoring and targeted therapies. Imagine a xenobot designed to navigate the human bloodstream. By tuning its chemical sensitivity, it could move toward areas with high concentrations of certain molecules, such as inflammatory markers, tumors, or pathogens.

This would allow xenobots to locate disease sites with unprecedented precision and potentially deliver targeted therapies without invasive procedures.

In environmental contexts, xenobots could play a role in pollution detection and mitigation. Their sensitivity to chemical gradients allows them to identify pollutants like heavy metals or toxins in water sources.

By swarming toward contaminants, xenobots could help localize pollution sources or even absorb toxins, offering a unique solution for environmental cleanup efforts. The biodegradable nature of xenobots ensures that they can perform these tasks without introducing long-lasting pollutants.

Challenges and Future Directions

While xenobots' sensing capabilities are groundbreaking, several challenges remain. Their current sensory range is limited, making it difficult to detect fine or distant stimuli compared to advanced electronic sensors. Researchers are working on ways to enhance xenobots' sensory precision, possibly by embedding specialized cells or modifying bioelectric fields to create more sophisticated responses.

Another challenge lies in ensuring controlled and predictable behavior. Although xenobots can navigate autonomously, guiding them precisely through complex environments remains a work in progress.

Future xenobots may incorporate engineered genetic or cellular modifications to enhance their responsiveness, expand their sensory repertoire, and increase their precision.

A New Horizon for Biohybrid Machines

Xenobots represent a new frontier where biological and technological boundaries merge. Their ability to sense and navigate environments based on cellular responses could revolutionize fields ranging from medicine to environmental science.

Unlike traditional robots, which rely on rigid sensors and programming, xenobots leverage life's inherent adaptability, setting the stage for biohybrid machines capable of complex tasks in unpredictable surroundings.

This journey to harness xenobots for autonomous sensing and navigation is still in its early days. Yet, as scientists deepen their understanding of cellular communication and bioelectric patterns, the potential to create even more sophisticated and responsive xenobots continues to grow.

It's an exciting chapter in the "Xenobot Chronicles," where sensing, navigating, and interacting with the world may become as natural for robots as it is for any living organism.

Chapter 8: Xenobot Brain: Neural Networks and Control

Imagine a biological robot with a brain—one that doesn't look like the traditional gray matter we associate with thought, but instead is an emergent network capable of responding to its environment, navigating, and even adapting to new challenges.

In the field of xenobiology, we are witnessing a revolution where biology and artificial intelligence intersect.

Xenobots, the world's first biological robots, represent the frontier of this intersection, especially when it comes to creating functional control systems.

While xenobots lack a central nervous system or traditional brain, their design relies on advanced neural networks and control mechanisms that mimic the brain-like processes of more complex organisms.

Building "Brains" in Xenobots

The "brain" of a xenobot operates on a combination of biological and computational principles. Traditional robots use an artificial neural network, which is a computational model inspired by the human brain's structure and processing patterns.

Neural networks excel in tasks requiring adaptation and learning by creating complex interconnections that adapt based on input.

In xenobots, scientists have created similar structures by influencing cell behavior, effectively teaching these organisms to interact with their surroundings in intelligent ways without any neurons.

To build a xenobot brain, scientists utilize frog embryo cells (Xenopus laevis) that exhibit basic response patterns, such as movement and clustering.

By using synthetic biology techniques, researchers mold these cells into specific configurations that can perform programmed tasks.

This form of "programming" relies not on digital code but on cellular communication pathways, where cells receive signals from each other and respond collectively.

Cells in the xenobot are carefully positioned so they form simple networks capable of self-coordination—somewhat similar to how neurons transmit signals across synapses.

Control Mechanisms in Xenobot Neural Networks

Without traditional neurons, xenobots require an alternative approach for controlled movement and decision-making. Scientists have discovered ways to create a simple feedback system that regulates their actions.

This feedback system involves specific cellular structures acting as sensors and actuators, so each xenobot can detect environmental cues and react appropriately.

The cells communicate through chemical signals, ensuring that each cell understands its "role" within the larger network.

For example, when a xenobot encounters a physical barrier, some of its cells can respond by sending biochemical signals to the other cells, prompting the entire structure to change direction.

This form of control, known as emergent behavior, allows xenobots to exhibit complex and seemingly purposeful movements without any direct programming.

Instead, control emerges from interactions among cells—a phenomenon sometimes referred to as "swarm intelligence." By fine-tuning these interactions, researchers can effectively direct xenobots to perform specific actions, such as moving toward a chemical signal or away from obstacles.

Programming Xenobot Neural Networks with AI Models

One of the most remarkable aspects of xenobot brain development involves integrating machine learning and artificial intelligence (AI) models. Researchers employ computational simulations to determine optimal cell configurations and movement patterns before assembling the biological xenobots. In these simulations, an AI model can predict how specific cell arrangements will behave under different environmental conditions, allowing researchers to pre-design xenobots for particular tasks, such as tissue repair or targeted drug delivery.

Machine learning algorithms help the xenobots learn how to navigate and perform tasks with higher efficiency. In some experiments, AI-driven models have evolved complex behaviors, including basic forms of self-repair when damaged and changes in movement patterns in response to new obstacles. By continuously feeding data from xenobot behavior back into the machine learning models, scientists can optimize neural control systems over time, creating xenobots that can adapt and improve their functionality.

Self-Organization and Adaptive Intelligence in Xenobots

The control systems in xenobots operate not as rigidly programmed routines but as adaptive, self-organizing networks. This adaptive intelligence, while limited compared to that of organisms with traditional brains, is crucial for the unique environments xenobots may encounter. Self-organization in xenobots is achieved by guiding how cells within the xenobot communicate and respond to each other, as well as how they react to external stimuli.

For example, a xenobot designed to seek out certain molecules in an environment will naturally "swim" toward a higher concentration of that molecule due to chemical gradients. Cells within the xenobot are preconditioned to recognize these molecules and move in response, creating a natural form of control that is simple but highly effective. This form of intelligence is adaptive because it doesn't require any central processing unit; instead, each cell contributes to the overall behavior, making the xenobot robust against changes in the environment and damage to individual cells.

The Future of Xenobot Brain Development: Toward Autonomy and Learning

As researchers continue to improve xenobot neural networks, the potential for developing autonomous and even learning-capable xenobots grows. Current experiments aim to create xenobots that can remember previous experiences, much like a simple form of memory, enabling them to make decisions based on past interactions with their environment. A xenobot with memory could, for instance, learn to avoid obstacles or remember the location of specific substances, making them more effective in search-and-repair applications.

Learning in xenobots can be implemented using bioelectric signaling. Bioelectricity is the natural voltage gradient that occurs across cellular membranes, and by modifying these gradients, scientists can change how xenobots interact with each other and their surroundings.

Experimental modifications to bioelectric signaling allow xenobots to retain information, providing a simple form of learning that does not require a traditional neural system.

This concept of "learning without neurons" could open doors to applications where xenobots autonomously navigate complex environments, making decisions based on past encounters and self-correcting their paths.

Ethical Considerations and the Broader Implications

With xenobot neural networks advancing rapidly, ethical discussions are essential. As these structures become more sophisticated, questions arise regarding the autonomy of living, synthetic organisms. Though xenobots lack consciousness, their ability to adapt and respond intelligently to stimuli places them in a unique ethical category.

What level of control should be permitted in designing cellular networks that mimic brain functions? How do we ensure xenobot technology is used for beneficial purposes, such as medical and environmental applications, rather than potential misuse?

Future xenobot brains may even integrate sensors to detect and respond to a broader range of signals, including biochemical markers of diseases or pollutants. Such applications have tremendous potential to revolutionize healthcare, environmental monitoring, and other fields.

However, the more autonomous and brain-like xenobots become, the more pressing it becomes to address questions of regulation, safety, and ethical responsibility.

The journey of creating a "brain" for xenobots exemplifies the fusion of biology and technology at its most innovative. By combining neural network principles with cellular behavior, scientists are pioneering a new kind of intelligence—one that operates without neurons, relying instead on the intrinsic capabilities of cells and the artificial intelligence models that guide them.

As xenobot neural networks continue to evolve, they promise to reshape not only how we approach biological robotics but also our understanding of intelligence, control, and the very concept of what it means to "think."

The xenobot brain is a small but significant step toward a future where living machines help address some of our most pressing challenges.

Simulating Neural Networks to Control Xenobot Behaviors

In the evolving world of bioengineering, the concept of xenobots has sparked fascination.

These tiny, living organisms, engineered from frog cells, are no longer just an idea; they are real, autonomous biological machines with the potential to change fields ranging from medicine to environmental science.

But what gives these living robots—xenobots—the ability to move, coordinate, and adapt?

A core part of their functional versatility is grounded in neural network simulations, which mimic brain-like processes to direct their behavior.

What Are Neural Networks, and Why Are They Important?

Neural networks are computer systems inspired by the human brain's structure and function. The brain operates through a complex network of neurons that communicate through electrical signals, allowing us to think, respond, and move. Neural networks in computing, or artificial neural networks (ANNs), seek to mimic this process, though they are digital and operate within computer code. These ANNs "learn" by adjusting their connections, which enables them to process inputs and produce specific, often complex, outputs.

Simulating neural networks to control xenobots is particularly effective because xenobots have no naturally occurring brain or nervous system. Through simulation, researchers essentially design a virtual "mind" that can be uploaded into xenobot biological architecture. This enables xenobots to execute specific tasks autonomously, adapting to environmental stimuli even without conscious thought or sensory organs. By embedding these networked behaviors, xenobots can navigate, move toward or away from targets, and even modify their paths in response to changes around them.

How Are Neural Networks Simulated?

Creating a neural network to control xenobots is a meticulous process that begins with designing an algorithm that can process inputs—data representing factors like environmental temperature, surface texture, or proximity to objects. Each layer of the network transforms these inputs, gradually refining the output that determines the xenobot's movement or response.

This process is largely managed through machine learning. The neural network is "trained" using vast datasets that allow it to adjust connections based on trial and error. For instance, suppose researchers want a xenobot to swim toward light. The network starts by generating random movement patterns. Through multiple iterations, it learns which patterns bring the xenobot closer to the light, adjusting until it reliably "swims" in the desired direction. Once trained, the network can then be embedded into xenobot systems, allowing the organisms to execute this behavior autonomously.

Reinforcement Learning: Training Xenobot Behaviors

One of the main techniques for training these neural networks is reinforcement learning. In this setup, the network learns by receiving feedback or "rewards" for successful actions and "penalties" for undesirable outcomes. For example, if a xenobot is programmed to avoid an obstacle, its neural network will receive a reward for every successful avoidance and a penalty if it collides with the obstacle. Over time, this feedback fine-tunes the network, encouraging the xenobot to avoid collisions reliably.

Reinforcement learning allows xenobots to adapt their actions according to specific goals. A xenobot equipped with a neural network trained through reinforcement learning could, theoretically, recognize harmful substances in water and actively avoid them, enhancing environmental applications like pollution detection and cleanup. By simulating thousands of scenarios, researchers create a neural network model that can be uploaded into xenobots, giving them the ability to respond intelligently and adaptively without human intervention.

Types of Behaviors Neural Networks Can Produce

Simulating neural networks for xenobots opens a broad spectrum of potential behaviors.

Some notable ones include:

1. Navigation and Pathfinding: Xenobots can be trained to move in specific directions or follow certain paths. This capability is crucial for tasks such as drug delivery, where xenobots may need to travel to a precise location within the body to release a therapeutic agent.

2. Environmental Sensing: Neural networks can help xenobots detect changes in their surroundings, like temperature shifts, light intensity, or the presence of chemicals. By adjusting behavior based on environmental feedback, xenobots can respond more dynamically to challenges, making them ideal for environmental monitoring or targeted interventions.

3. Collective Behavior and Coordination: When xenobots are placed in groups, their neural networks can be trained to facilitate collective actions, such as forming specific patterns or moving in unison. These behaviors could allow xenobots to complete tasks requiring collaboration, such as pushing objects or clearing blocked passages, and could have applications in medical or disaster response settings.

4. Self-Repair and Adaptive Resilience: Advanced neural networks are even beginning to enable limited self-repair. Xenobots can respond to damage by adapting their structure or redirecting cellular functions to mitigate injury.

This adaptability could prove especially useful for creating robust xenobots that can endure in hostile environments.

Challenges and Limitations of Simulating Neural Networks in Xenobots

While the use of neural networks to control xenobot behavior is groundbreaking, several challenges remain. For one, biological systems are inherently unpredictable. Neural networks operating in silicon are very different from their counterparts in biological xenobots, which are susceptible to environmental influences, cell degradation, and other organic variables.

Additionally, creating more complex, precise behaviors requires significant computational resources. Simulating millions of neural interactions to achieve detailed, lifelike responses is a formidable task.

Although strides in computing power have made it possible, the resources needed to simulate highly specialized behaviors are still a limiting factor.

Ethical considerations also come into play. As neural networks in xenobots advance, questions about control and autonomy arise. How much autonomy should a xenobot possess? Should there be limitations on certain behaviors? Ensuring that xenobot neural networks remain within ethical boundaries will be crucial as the technology progresses.

Future Prospects: Toward Enhanced Intelligence in Xenobots

As research on neural networks and xenobot behavior continues, we can expect even more sophisticated, responsive xenobots capable of performing complex tasks autonomously. Scientists are exploring hybrid models that combine neural networks with other computational approaches, like fuzzy logic, to improve xenobot adaptability and efficiency in uncertain or dynamic environments.

Moreover, advancements in neurobiological engineering are pushing toward real-time responsiveness in xenobots, enabling them to adjust their behavior instantly in unpredictable situations.

For instance, a xenobot tasked with detecting and isolating specific types of bacteria could soon adapt its behavior based on real-time analysis of bacterial markers, transforming how we approach infection control and environmental decontamination.

In the not-so-distant future, neural networks may allow xenobots to "learn" continuously, similar to how animals develop skills over time.

Such capabilities would make xenobots invaluable for applications that require ongoing adaptation, like long-term environmental monitoring, where conditions are constantly changing.

The Intelligence of Tomorrow, Today

Simulating neural networks for xenobot control is an incredible stride toward creating lifeforms with engineered intelligence.

By embedding neural network models into these biological robots, scientists are enabling them to perceive, react, and even adapt to their environment, expanding the boundaries of robotics and bioengineering.

With ongoing advancements, xenobots are set to redefine our approach to medicine, environmental science, and beyond, heralding a future where biologically-driven artificial intelligence may soon become an integral part of solving the world's most pressing challenges.

Achieving Complex Decision-Making Using Artificial Intelligence

The ability to make complex decisions has long been the hallmark of higher intelligence in both biological organisms and machines. In the realm of robotics, decision-making has traditionally relied on programming, where algorithms determine responses based on pre-set conditions. But as we begin to explore and harness the potential of xenobots—living, programmable organisms crafted from biological tissues—the integration of artificial intelligence (AI) into these tiny entities opens doors to decision-making processes previously unseen in nature or conventional robotics.

Understanding Decision-Making in Biological Contexts

In nature, decision-making evolved as a survival mechanism. For example, animals decide when to move, hunt, or hide based on sensory inputs and internal drives. These decisions are informed by complex neural networks that filter, prioritize, and respond to a dynamic environment. Likewise, human decision-making relies on a blend of instinct, experience, memory, and prediction. Mimicking these nuanced decision processes in xenobots, however, presents unique challenges and opportunities.

Biological tissues, such as the cells that form xenobots, naturally respond to stimuli like chemical signals or physical contact, enabling some form of rudimentary decision-making. In the absence of a nervous system, though, how can xenobots perform more complex tasks? This is where AI and computational models come into play, providing xenobots with an additional layer of "thought" or response patterns that mimic intelligent decision-making.

The Role of Artificial Intelligence in Xenobot Decision-Making

AI is a driving force behind the development of decision-making in xenobots. Through machine learning algorithms, xenobots can analyze environmental conditions and adjust their behaviors accordingly. Here, AI enables xenobots to engage in activities beyond simple reactions, giving them the potential to process multiple stimuli and "choose" optimal responses.

One of the most promising applications of AI in xenobots involves reinforcement learning, a subset of machine learning where AI algorithms "learn" from feedback. When xenobots are placed in various environments, their AI programming allows them to try different actions and receive feedback in the form of success or failure, reinforcing behaviors that produce positive outcomes.

This learning process is akin to how animals learn to navigate their surroundings and find food or shelter.

Another essential aspect of AI integration in xenobots is pattern recognition. AI allows xenobots to identify and categorize stimuli, recognizing, for instance, different types of surfaces or the presence of specific chemicals.

Through continuous exposure and feedback, xenobots learn to distinguish between friendly or threatening stimuli, which could be crucial for applications like environmental monitoring or medical therapies.

For example, a xenobot tasked with locating cancerous cells may learn to differentiate these cells from healthy ones through repeated encounters, sharpening its decision-making accuracy over time.

The Mechanisms Behind AI-Driven Decision-Making

The backbone of AI-driven decision-making in xenobots lies in neural networks—computational structures that mimic the way neurons function in the brain.

While xenobots do not have traditional neural networks, researchers have used artificial neural networks (ANNs) to simulate decision-making paths in these tiny biological robots.

These networks can be "trained" to handle complex scenarios by adjusting weights and biases, elements that influence how data is processed within the network. Through this process, ANNs help xenobots determine the best action to take based on environmental conditions.

For example, let's imagine a xenobot equipped with AI to navigate a maze. The neural network within its AI would process sensory information, learning which turns lead to dead ends and which lead closer to the goal.

Over multiple trials, the xenobot would "learn" the optimal path, allowing it to make accurate decisions about which direction to turn, even when the environment changes slightly. This type of decision-making, while seemingly simple, is a significant leap from basic, pre-programmed responses, illustrating the power of AI in guiding xenobot behavior.

Challenges in Achieving Complex Decision-Making

Despite these advancements, integrating complex decision-making into xenobots presents significant challenges. The most notable issue is the lack of conventional sensory systems in xenobots. Unlike robots with electronic sensors, xenobots rely on biological responses to stimuli, which are more limited and less precise. Researchers are actively exploring ways to enhance sensory capabilities through chemical or genetic modifications, allowing xenobots to detect more environmental cues and, consequently, make better-informed decisions.

Additionally, there is the challenge of scalability. AI algorithms can be computationally intensive, and xenobots are often microscopic. Packing sophisticated AI capabilities into such small biological structures requires highly efficient algorithms that use minimal resources.

Innovations in edge AI—a field focused on processing data locally on small devices without relying on powerful remote servers—are crucial to overcoming this hurdle.

Another challenge involves maintaining ethical and safety standards. As xenobots become more capable of autonomous decision-making, questions about control and potential unintended consequences arise. Scientists are carefully considering safety measures and ethical guidelines to prevent xenobots from making decisions that could harm their environment or any living beings.

Real-World Applications and the Future of AI-Driven Xenobots

The potential applications of AI-driven decision-making in xenobots span multiple fields. In medicine, xenobots could one day be programmed to navigate through the human body, identifying and eliminating harmful cells or delivering targeted treatments. For instance, xenobots might recognize and "decide" to attach to a cancer cell and release a specific therapeutic agent, sparing healthy cells and improving treatment precision.In environmental sciences, xenobots could be deployed to monitor ecosystems, detect pollutants, or clean up microplastics in water bodies. Their AI-driven decision-making would enable them to adapt to dynamic and complex environments, making real-time adjustments as they encounter various pollutants or obstacles.

Looking to the future, researchers envision even more ambitious roles for xenobots. With advances in AI, xenobots may one day be able to "collaborate" with one another, coordinating actions and sharing information to achieve common goals. Imagine a swarm of xenobots collectively assessing an environment, pooling their observations to make informed decisions, much like how a bee colony communicates to find food sources.

The Road Ahead

While xenobots are still in their early stages, the integration of AI-driven decision-making signals a paradigm shift. By blending biology with advanced computing, scientists are pioneering a new class of living machines capable of autonomously navigating and adapting to the complexities of the real world. These AI-enhanced xenobots open up a realm of possibilities that challenge the traditional boundaries between the biological and the technological, suggesting that, in the not-so-distant future, the line between natural and artificial intelligence may blur even further.

In the unfolding journey of xenobots, achieving complex decision-making through AI is a critical milestone. It brings us closer to a future where living, programmable organisms perform sophisticated tasks, potentially transforming fields like medicine, environmental science, and beyond.

Chapter 9: Xenobots as Environmental Explorers

In recent years, a new type of biological machine has captured the imagination of scientists and environmentalists alike: the Xenobot. These tiny, programmable, living robots, derived from the stem cells of the African clawed frog (Xenopus laevis), offer unprecedented potential to explore and assist in safeguarding our planet's ecosystems.

With their unique capabilities, Xenobots stand out as potential allies in environmental monitoring and restoration—a field where their combination of biological adaptability and programmable behavior can make a measurable impact.

1. Xenobots Explauned

Xenobots are small, bioengineered structures created from the skin and heart cells of frogs. Despite their microscopic size, they have unique, emergent properties that make them distinct from other forms of life and machine. The Xenobot's skin cells provide structural support, while heart cells are used to create movement, thanks to their inherent ability to contract.

This combination allows scientists to "program" Xenobots to move, interact with each other, and perform specific tasks. Unlike traditional robots, they do not have any metal, electronics, or plastics, and because they are biodegradable, they won't leave harmful traces in the environment.

2. Why Xenobots for Environmental Exploration?

Xenobots possess certain advantages that make them uniquely suited to environmental applications:

- **Biodegradability:** Made entirely from biological material, Xenobots naturally decompose once their task is complete, avoiding the environmental concerns associated with plastic or metal pollution.

- **Self-Healing:** Minor damage to a Xenobot can be repaired as its cells naturally regenerate. This ability to self-heal makes them durable and efficient for long-term environmental projects.

- **Biocompatibility:** Since Xenobots are created from living tissue, they are less likely to disrupt ecosystems compared to traditional robots.

Their biocompatibility means they can interact with the environment in more natural ways, moving around and through obstacles like soil, sediment, or plant roots.

- **Programmable Movement:** By designing different shapes or patterns of Xenobots, researchers can create structures that move in specific ways, making them capable of navigating complex environmental terrain.

3. Xenobots in Pollution Detection and Clean-Up

One of the most promising applications of Xenobots is their potential to help detect and mitigate pollution. Currently, pollutants such as microplastics and toxic chemicals present major challenges to ecosystems around the globe.

Xenobots could become invaluable tools in detecting, identifying, and even neutralizing harmful contaminants.

Tracking Microplastics

Microplastics have become pervasive pollutants in water bodies, causing harm to aquatic life and entering the human food chain. Xenobots could be designed to locate and gather microplastics in aquatic environments, "sweeping" through areas where other cleanup devices might not reach.

They could also transport microplastics to collection points or aggregation sites, where they could be removed more effectively.

Using sensors or chemical markers, Xenobots could even distinguish microplastics from organic debris, improving the accuracy and efficiency of cleanup efforts.

Bio-remediation of Chemical Spills

Chemical pollutants from industries and agriculture often find their way into rivers, oceans, and soil, harming ecosystems. Certain types of Xenobots could potentially carry enzymes that neutralize or break down specific toxins, reducing environmental harm. Imagine a swarm of Xenobots programmed to locate areas contaminated with oil, pesticides, or heavy metals and release bioactive substances to help neutralize these contaminants.

As biological organisms, Xenobots could be safer alternatives to more invasive cleanup methods that may inadvertently harm the surrounding ecosystem.

4. Monitoring Environmental Health

Xenobots could also be deployed to monitor the health of environments over time. Equipped with sensors or biochemical markers, they could measure indicators like pH levels, oxygen content, and temperature.

With their small size and mobility, Xenobots could reach areas that traditional monitoring equipment cannot access, such as the undersides of rocks in rivers or the roots of plants in soil.

Assessing Water Quality

Water quality is essential to the health of aquatic ecosystems and human communities. Traditional water monitoring systems require significant infrastructure and can be costly, especially in remote or underdeveloped areas. Xenobots could provide a flexible, low-cost alternative.

For example, they could be programmed to collect data on nutrient levels or contaminant presence, providing real-time information about water quality.

Their movements could even be coordinated to sample a wide range of points within a body of water, creating a dynamic map of environmental health.

Soil Health Assessment

In agriculture and natural ecosystems alike, soil health is a critical factor for plant growth, biodiversity, and carbon sequestration. Xenobots could serve as a new method for soil monitoring, navigating through various layers of soil and gathering data on moisture levels, pH, and nutrient composition. This information could be invaluable for both farmers and conservationists, helping to optimize soil management practices that promote sustainable agriculture and ecosystem resilience.

5. Aid in Ecosystem Restoration

Ecosystem restoration is a high priority in conservation, and Xenobots have the potential to play a part here as well. Whether aiding in the planting of new vegetation, distributing beneficial bacteria, or even transporting seeds, Xenobots could contribute directly to rewilding and habitat restoration.

Seed Dispersion and Reforestation

In reforestation projects, Xenobots could be programmed to disperse seeds in targeted areas, particularly those that are difficult to reach by humans or traditional machinery. By mimicking natural dispersal agents like animals or wind, Xenobots could help plant seeds at optimal distances and depths, promoting healthy forest growth.

They could even carry nutrient-rich soil or fertilizer to support young plants, increasing their chances of survival in challenging environments.

Coral Reef Restoration

Coral reefs are among the most fragile ecosystems, and they're particularly challenging to restore due to their sensitivity to temperature changes and pollutants. Xenobots could play a role in coral reef restoration by delivering coral larvae to affected areas, or even applying probiotics to corals to enhance their resilience to disease.

As biodegradable and biocompatible tools, Xenobots could operate in reef environments with minimal risk of disrupting marine life.

6. Challenges and Future Directions

While the potential for Xenobots as environmental explorers is immense, several challenges must be overcome before widespread deployment. Currently, researchers are working on enhancing Xenobots' sensing capabilities, expanding their programming complexity, and improving their energy efficiency.

Additionally, there are ethical and regulatory considerations, as releasing living, programmable robots into natural environments poses questions about containment and impact.

However, with continued advancements, Xenobots could transform how we approach environmental stewardship, enabling new, sustainable strategies to monitor, protect, and restore ecosystems. As a new class of bio-inspired technology, they offer a glimpse into a future where living machines and nature work together in harmony to address some of the planet's most pressing environmental challenges.

With each new development, the potential for Xenobots as environmental explorers becomes more tangible, opening doors to a cleaner, more sustainable world.

Applications of Xenobots in Exploring Hazardous Environments

In recent years, a fascinating breakthrough in biotechnology has captured the imagination of scientists and the public alike: xenobots.

These programmable, self-healing biological robots, derived from frog stem cells, are engineered to perform tasks that challenge traditional robotics.

One of the most exciting and impactful applications of xenobots lies in their potential to explore hazardous environments.

These micro-bots could transform the way we investigate and monitor environments that are otherwise too dangerous, inaccessible, or delicate for human intervention or traditional robots.

1. Understanding Xenobots and Their Unique Abilities

Xenobots are not your typical machine-based robots. They are clusters of living cells, specifically engineered using cells from Xenopus laevis, a species of African clawed frog.

Scientists at Tufts University and the University of Vermont pioneered xenobots by shaping these cells into small, mobile, and programmable forms.

By controlling the growth and pattern of these cells, researchers have created small biological robots capable of simple movements, self-repair, and responsiveness to their surroundings.

The biological nature of xenobots gives them several advantages over conventional robotic systems, especially in hostile environments.

Xenobots can potentially heal themselves after minor damage, require minimal energy to operate, and degrade naturally, leaving no toxic waste.

Additionally, they are small enough to reach places conventional robots might struggle with, such as the tiny fissures in rocks or the narrow underwater crevices in coral reefs. These traits make xenobots particularly well-suited for hazardous explorations.

2. Applications in Extreme and Hazardous Environments

Xenobots' resilience and adaptability present a wide range of possibilities for exploring dangerous settings, where other robotic solutions would either fail or require extensive modification to operate.

Some of these challenging environments include radioactive zones, polluted waters, and areas at risk for biological or chemical contamination.

By deploying xenobots in these settings, we can gather valuable data, monitor hazardous materials, and even potentially aid in environmental cleanup efforts.

a) Radiation Zones

One of the most compelling uses of xenobots is in the exploration of radioactive areas, such as the containment zones around nuclear disaster sites like Chernobyl and Fukushima. Radiation poses severe risks to both humans and traditional machinery, often causing equipment to degrade over time. Xenobots, however, could provide a way to navigate these areas without the typical structural breakdown seen in mechanical devices. Since xenobots are made of organic material and can regenerate, they may last longer than metallic or plastic robots in environments where radiation corrodes materials quickly.

Xenobots could be used to monitor radiation levels in real time by navigating specific areas, collecting samples, and relaying data to a safe location. This information could help scientists understand the spread of radiation, detect leaks, and even assess the effectiveness of containment measures in contaminated zones.

b) Underwater and Deep-Sea Exploration

The ocean's depths are another hazardous frontier where xenobots could make substantial contributions. Traditional underwater robots are often limited by their bulk, high energy consumption, and vulnerability to pressure changes. Xenobots' small size, low energy needs, and organic nature allow them to move more seamlessly through delicate ecosystems such as coral reefs or complex underwater cave systems. They could be designed to monitor marine life health, assess coral reef damage, and detect microplastic pollution in these environments without causing damage or disturbing ecosystems.

Xenobots also offer a unique potential for exploring deep-sea environments that are usually hostile to human activity. These organisms could function autonomously at various depths, gathering information about undersea vents, unique microbial life forms, and unknown species that thrive in the darkness of the ocean floor. Because xenobots could be programmed to operate independently or in swarms, they might even be able to work together to map complex environments, sharing data across units to create a comprehensive picture of the area being explored.

c) Pollution and Contaminated Waterways

Another promising application of xenobots lies in polluted and contaminated waters. Traditional robots and drones face several limitations in these settings, especially when pollutants include hazardous chemicals or heavy metals that can damage electronic systems. Xenobots, being organic and biodegradable, offer a distinct advantage in these scenarios. They could not only navigate polluted waterways to monitor contamination levels but could potentially absorb and contain harmful substances, acting as mobile biofilters to aid in remediation efforts.

Additionally, xenobots could monitor bacterial or toxic waste, providing insights into the extent and type of contamination present. For instance, in oil spill scenarios, xenobots could assess the damage and help track the spread of pollutants.

Their ability to carry and deliver small payloads might also be harnessed to deploy chemicals that neutralize toxins or encourage the breakdown of harmful substances in water bodies, aiding in active cleanup efforts.

d) Search and Rescue in Dangerous Terrain

Xenobots' mobility and size also make them promising tools for search and rescue operations in treacherous terrains such as collapsed buildings, mine shafts, and avalanche zones. Often, survivors in these environments may be trapped in pockets of air inaccessible to larger rescue equipment. Swarms of xenobots could be deployed to locate these pockets, carrying sensors to detect signs of life, including temperature changes and CO_2 levels.

They could potentially carry small emergency supplies, such as water or nutrients, to sustain trapped individuals until larger rescue operations can be mobilized.

Moreover, xenobots can reach places where traditional robots would struggle due to debris, unstable structures, or limited access. By relaying data back to rescue teams, xenobots could provide valuable information on the survivors' location and surrounding conditions, enabling more effective planning for rescue operations.

3. Ethical and Environmental Considerations

While xenobots offer groundbreaking possibilities for exploring and monitoring hazardous environments, there are ethical and ecological considerations associated with their deployment. Being biodegradable means xenobots are less likely to cause pollution, but their release into sensitive ecosystems could disrupt local biology. Researchers are actively working on protocols to ensure that xenobots are programmed to degrade completely after their tasks, minimizing any potential environmental impact.

The use of xenobots also raises questions about control and oversight. Ensuring they cannot evolve or act beyond their intended programming is critical, especially in ecosystems where unintended actions could have significant repercussions. Advances in biocompatibility and the strict regulatory standards governing their deployment will be essential as xenobots transition from experimental to practical applications.

Xenobots represent a paradigm shift in how we might explore and interact with hazardous environments. Their unique biological makeup enables them to operate in settings that are challenging for traditional robots or dangerous for humans. As the technology matures, xenobots may become invaluable tools for monitoring, cleanup, and exploration, offering safe, sustainable, and efficient solutions for navigating the Earth's most hostile landscapes. In the future, they may even allow us to explore places previously beyond reach, ushering in a new era of environmental interaction and discovery.

Xenobot Missions in Search and Rescue Scenarios

Imagine a tiny, intelligent creature, small enough to navigate through rubble and strong enough to withstand challenging environments, searching tirelessly for signs of life. In the unfolding saga of biological robotics, xenobots are some of the most promising agents for tasks like these, specifically in scenarios where traditional machines fall short. These tiny, programmable bio-robots offer a unique combination of biological adaptability and robotic control, making them uniquely suited for high-stakes, high-precision search and rescue (SAR) missions.

Xenobots, designed by assembling frog cells into motile, controllable biological forms, are living machines with a wide range of potential applications. What makes them so suitable for SAR is their ability to move through small, complex environments, detect biological signals, and perform programmed actions autonomously or in response to environmental cues. Let's look at some example missions where xenobots could be deployed, showcasing their immense potential to save lives in dangerous situations.

1. Search and Rescue in Collapsed Buildings

When earthquakes or other disasters strike, collapsed buildings present one of the most dangerous environments for rescue workers. Humans and even robots often struggle to navigate these unstable, confined spaces. Xenobots, however, are tiny and flexible enough to move through the smallest cracks. Imagine a scenario where a xenobot swarm, composed of hundreds of tiny units, is deployed to find survivors trapped under debris.

Each xenobot in this scenario would be programmed to sense certain environmental cues, such as body heat, carbon dioxide levels, or chemical traces associated with human respiration. Upon detecting these signals, the xenobots would cluster together and "flag" the location, allowing human rescuers to focus their efforts more efficiently.

Because they're made from biological material, xenobots can safely operate in delicate environments without posing the risk of sparks or electrical interference—a critical advantage in disaster zones where gas leaks could make traditional robotics unsafe.

2. Flooded Area Exploration and Victim Location

Flooded areas present another challenge where xenobots could excel. Floodwaters are often murky, cluttered with debris, and present severe hazards to traditional machines due to the high levels of water and debris interference.

Xenobots, however, could swim through these waters, moving in search of survivors or mapping underwater hazards.

Imagine a flood scenario where people are trapped in partially submerged buildings or cars. Rescuers struggle with low visibility and can't easily assess the exact number and location of people in distress. Here, xenobots can move through water to detect signals like vibrations (from movement or banging) and even sound waves produced by survivors. They could work as a team to relay information back to rescuers, even creating pathways to facilitate human entry and aid.

In cases where survivors are trapped in vehicles, xenobots could detect the specific sounds associated with human life, such as tapping or vocal sounds. Their ability to navigate tight spaces underwater and move toward signs of human presence could be life-saving in such scenarios.

With an option to emit a faint bioluminescent glow, these xenobots could help rescuers by illuminating areas or signaling their location for easy identification.

3. Detection of Hazardous Materials in Industrial Accidents

In many industrial disasters, hazardous chemicals are often released into the air or groundwater. Xenobots can be programmed to detect specific chemicals, like ammonia or benzene, often present in industrial environments. They can help in monitoring and identifying these chemicals without risking human lives.

In the aftermath of an industrial accident involving hazardous substances, xenobots could play a critical role in both identifying and locating survivors and monitoring environmental hazards. Consider a chemical plant explosion where airborne toxins spread across a wide area.

Xenobots could be introduced into the environment to seek out areas with high concentrations of specific toxic chemicals, mapping their presence while also detecting life signs from survivors who might have been incapacitated by the fumes.

With data from these xenobots, human rescuers would know where to take protective measures and focus their efforts, potentially saving lives while reducing the risk to rescue teams.

4. Fire Search Missions in Smoke-filled Environments

In the event of a fire, one of the biggest dangers is the thick smoke that severely limits visibility and makes it difficult to locate individuals in need of help. Xenobots are unaffected by smoke and could quickly navigate through it, equipped with sensors to detect the heat signature of a trapped person.

In this type of mission, a swarm of xenobots could be sent into a burning building to fan out and locate survivors. These xenobots could work collaboratively, communicating heat locations to each other and moving toward the strongest thermal signals that may indicate human presence. Unlike drones or robots, xenobots are much smaller and can penetrate tighter areas, allowing them to reach spaces where someone might be taking refuge to escape the flames.

Xenobots could even carry tiny sensors or compounds capable of detecting human exhalation patterns or unique markers of living tissue. Once they locate a survivor, they could emit a signal—such as a fluorescent glow—to help guide rescuers directly to the person. This unique ability to navigate and identify survivors in the smoky, hot environment of a fire-affected area makes xenobots a game-changer in fire rescue missions.

5. Landslide or Avalanche Victim Detection

Landslides and avalanches are some of the most challenging SAR scenarios due to the heavy, dense layers of earth or snow that cover victims. Traditional robots often have difficulty penetrating deep enough to detect life, while sending human rescuers can be risky due to potential secondary slides or shifts.

Xenobots could tackle this by working in large numbers and burrowing through loose soil or snow. Equipped with biological sensors to detect CO_2 levels—a reliable indicator of human respiration—these xenobots could move through the layers, mapping potential locations of trapped individuals. By creating "marker" zones where they detect signs of life, xenobots could guide rescue teams to the most promising areas.

In avalanche scenarios, xenobots could navigate the compact snow, moving through spaces and burrowing with minimal disturbance, reducing the risk of further snow shifts. If they find trapped individuals, they could again signal their position through bioluminescent markers, allowing for a quick response.

6. Mine Rescue Operations

In mining accidents, survivors can be trapped in narrow, dark spaces with unstable structures, making it dangerous for large machinery or human rescuers to access. Here, xenobots could excel due to their small size and flexible mobility.

Deployed into the mine, xenobots would scan for signs of life through heat, carbon dioxide, or sound detection. They could explore cracks and narrow passages that are inaccessible to larger robots.

By forming "breadcrumbs" or signals along the way, they could map out a path for rescuers, enabling a systematic and safe approach to reach the survivors.

The Future Potential

While still in their early stages, xenobot technology promises to revolutionize SAR missions. With continued advances, xenobots will likely become even more sensitive to environmental cues and adaptable in various terrains.

The power of biological robotics holds incredible promise for lifesaving missions in conditions that are too hazardous, confined, or complex for traditional machines, marking a new frontier in the technology of human aid and survival.

Chapter 10: Xenobots in Medicine and Healthcare

The Dawn of Living Machines

The field of medicine has always benefited from groundbreaking technologies, and xenobots—a new type of living organism crafted through biological engineering—represent one of the latest advancements with immense potential to revolutionize healthcare. Xenobots are tiny, programmable organisms built from frog cells (specifically Xenopus laevis, hence the name) and designed to perform specific tasks in controlled environments. Their unique properties, a blend of biology and technology, open doors to applications previously confined to science fiction.

Let's explore how xenobots might transform medicine and healthcare, from targeted drug delivery to tissue repair and diagnostics.

Xenobots are not traditional robots made of metal and circuits; instead, they are composed of living cells and are engineered to carry out functions guided by their cellular programming. These organisms are built by taking skin and heart cells from frog embryos and arranging them in configurations that allow them to move, cluster, and even "self-heal" when damaged. Heart cells are particularly useful because their rhythmic contractions enable motion without additional power sources, while skin cells provide structural support.

Using advanced algorithms and machine learning, researchers can determine the optimal shape and behavior for xenobots based on the specific tasks they are designed to perform. For example, xenobots can be shaped in such a way that they swim through liquid environments, making them ideal candidates for internal medical applications.

Their biodegradable nature is another advantage; since they're made from organic material, they break down naturally in the body after fulfilling their purpose, minimizing the risk of adverse reactions.

Applications in Targeted Drug Delivery

One of the most promising applications of xenobots in medicine is their potential use for targeted drug delivery. Traditionally, drug delivery within the body can be challenging, especially when trying to reach specific cells or tissues without affecting surrounding healthy areas. Xenobots offer an innovative approach to this problem by delivering medications directly to targeted sites.

Due to their small size and customizable mobility, xenobots can be directed to navigate through complex biological environments like the bloodstream or gastrointestinal tract. They could be engineered to carry small amounts of drugs and release them in controlled bursts when they reach a specific location. This controlled release would allow for more effective treatment with fewer side effects, as the drugs could act on diseased cells directly while minimizing exposure to healthy tissues.

For example, in cancer treatment, xenobots could potentially deliver chemotherapeutic agents directly to tumor cells, reducing the harm to surrounding healthy cells and minimizing the common side effects of chemotherapy, such as hair loss and immune suppression. Additionally, because xenobots degrade naturally, there would be no need for follow-up procedures to remove them from the body.

Tissue Repair and Regeneration

Another exciting application of xenobots in healthcare is their potential for tissue repair and regeneration. Xenobots have been shown to exhibit some degree of "self-healing" when damaged. Researchers believe this ability could be harnessed for repairing injuries or promoting tissue regeneration in humans.

Imagine xenobots programmed to move to the site of an injury and form a scaffold over damaged tissues. By assembling at the injury site, they could facilitate the healing process by creating a supportive structure that promotes cellular growth and regeneration. This application could be particularly beneficial for wounds that are difficult to heal, such as chronic ulcers or injuries resulting from diseases like diabetes.

In cases of severe trauma where tissues are too damaged for traditional healing, xenobots could be employed as temporary biological "patches" that support the healing process until the body regenerates the necessary tissue. This could be especially valuable in emergency medicine or battlefield scenarios, where immediate intervention is critical for survival.

Diagnostics and Early Disease Detection

Early detection of diseases often leads to better treatment outcomes, but traditional diagnostic tools can sometimes fail to identify issues at the cellular or molecular level. Xenobots, however, can be programmed to act as sentinels that patrol the body, detecting molecular markers or other signs of disease in real-time.

For instance, xenobots could be designed to change shape, emit light, or release specific biomarkers upon encountering particular proteins associated with diseases such as cancer or autoimmune disorders.

By monitoring these responses, healthcare providers could gain early warning signs of potential health issues, enabling faster diagnosis and treatment.

In a similar way, xenobots could help detect toxic substances or inflammatory markers in the body. For individuals with chronic conditions like rheumatoid arthritis or multiple sclerosis, xenobots might provide real-time monitoring of inflammatory markers, helping doctors tailor treatments to the patient's immediate needs and improving quality of life.

Combating Infectious Diseases

With infectious diseases continuing to pose significant health risks worldwide, xenobots could play a role in both prevention and treatment. Researchers envision xenobots that can patrol the body for harmful bacteria, viruses, or other pathogens.

Once detected, these xenobots could either directly neutralize the pathogens or release antibacterial or antiviral compounds in response.

This approach could be particularly useful for individuals with weakened immune systems or in cases where antibiotics and antivirals have proven less effective.

By deploying xenobots, it might be possible to control infections more precisely without the broad, often damaging effects of systemic drugs.

Additionally, xenobots could potentially serve as tools in epidemiology and public health monitoring, detecting pathogens in bodily fluids or even wastewater to give early warnings of disease outbreaks.

Ethical and Safety Considerations

As with any emerging technology, the development of xenobots raises ethical and safety questions. Since xenobots are technically living organisms, their use prompts questions about autonomy, environmental impact, and unintended biological interactions.

Researchers are closely studying these issues, as responsible development of xenobots is crucial to prevent potential harm.

There are also questions about the long-term interactions between xenobots and the human body. While current designs degrade within days, future versions of xenobots that last longer or have more complex functions may require more rigorous safety testing to ensure they don't trigger immune reactions or interfere with natural biological processes.

The Future of Xenobots in Healthcare

Xenobots represent an early but exciting foray into the concept of living machines—biological entities that blur the line between organism and technology. While still in the research phase, xenobots have already demonstrated capabilities that could transform medicine and healthcare in profound ways.

From targeted drug delivery and tissue repair to diagnostics and infection control, these living machines may soon become integral tools in medical science.

The future of xenobots in healthcare may include increasingly complex designs, greater functionality, and integration with other advanced technologies like artificial intelligence and nanotechnology. As the research progresses, xenobots could become a key component of personalized medicine, offering treatments tailored to each individual's unique biological makeup.

With responsible development, xenobots could help us usher in a new era in medicine, where living machines work alongside doctors and nurses, making healthcare more effective, targeted, and compassionate than ever before. The journey of xenobots in medicine has just begun, and their potential is as vast as it is exciting.

The Potential Role of Xenobots in Drug Delivery and Targeted Therapy

The field of drug delivery and targeted therapy is witnessing an unprecedented transformation with the emergence of xenobots, a new class of microscopic, biologically-engineered robots. Created from the stem cells of the African clawed frog (Xenopus laevis), these living organisms are designed to perform specific tasks at microscopic levels.

In drug delivery and targeted therapy, xenobots hold the potential to revolutionize how medicines reach specific areas of the body, offering highly targeted, controlled, and efficient treatment solutions.

The Birth of Xenobots: A New Frontier in Biomedical Science

Xenobots represent a blend of robotics and cellular biology. Developed by a team of biologists and computer scientists, these "living machines" are designed by carefully arranging and modifying frog stem cells to form small, programmed clusters. Xenobots are not robots in the traditional sense; they are more like programmable tissue with specific behaviors, such as movement, self-repair, and even replication.

Their small size, soft-body construction, and biological composition allow them to navigate bodily fluids and tissues with a level of ease and precision that conventional drug delivery systems cannot achieve.

Advantages of Xenobots in Drug Delivery and Therapy

1. Biodegradability and Biocompatibility

Since xenobots are built from organic cells, they are biodegradable and naturally compatible with the human body. Traditional drug delivery methods, like polymer-based nanocarriers, sometimes provoke immune reactions or leave residues within the body.

Xenobots, however, are made from entirely organic material and degrade into harmless biological waste once their task is complete. This property minimizes side effects and the risk of complications, particularly in long-term treatments.

2. Precision and Targeting Ability

One of the most significant limitations in drug therapy is getting the medication precisely where it is needed, especially in conditions like cancer or diseases confined to specific organs. Xenobots can be engineered to move purposefully toward particular sites within the body, directed by chemical gradients or navigated externally via magnetic fields.

This precise movement allows for direct and controlled drug delivery to target cells or tissues, reducing off-target effects and increasing the efficacy of treatment.

3. Self-Healing and Adaptability

Xenobots exhibit an extraordinary ability to repair themselves if damaged, making them particularly resilient in the challenging environments within the human body.

They can adapt to different conditions, navigating through complex biological structures such as blood vessels or tissue layers.

In drug delivery applications, this adaptability means that xenobots can withstand various physiological conditions without losing functionality, enhancing the reliability of the therapy they are administering.

4. Controlled Release Mechanism

Xenobots can be engineered to carry drugs within their cellular structure or on their surface, releasing them in response to specific triggers like pH changes, temperature variations, or chemical signals. This controlled release mechanism allows for a highly personalized treatment experience, as drugs can be delivered exactly when and where they are needed, maintaining therapeutic concentrations and minimizing the risk of side effects.

Applications in Treating Diseases with Precision Medicine

The use of xenobots in drug delivery could have transformative effects on several disease areas, particularly in the realm of precision medicine. Conditions such as cancer, neurodegenerative diseases, cardiovascular illnesses, and localized infections stand to benefit greatly from xenobot-based therapies.

1. Cancer Therapy

One of the most promising applications of xenobots is in cancer therapy, where precision and minimal off-target effects are essential. Xenobots could be directed to infiltrate cancerous tumors, delivering chemotherapy agents directly to cancer cells while sparing healthy cells. Additionally, xenobots could be programmed to respond to the acidic environment around tumors, releasing their drug load only when they reach malignant tissues. This approach could reduce the systemic side effects typically associated with chemotherapy.

2. Neurological Diseases

Treating neurological diseases poses a unique challenge due to the blood-brain barrier, which prevents many drugs from reaching the brain. Xenobots may offer a solution by crossing or navigating around this barrier to deliver drugs directly into brain tissues. For neurodegenerative diseases like Alzheimer's or Parkinson's, xenobots could be programmed to release neuroprotective agents, antioxidants, or growth factors to specific brain regions, potentially slowing disease progression and enhancing patients' quality of life.

3. Cardiovascular Therapy

Xenobots could also play a role in treating cardiovascular conditions by delivering drugs directly to damaged blood vessels or heart tissues. They could be directed to sites of atherosclerotic plaques to release drugs that dissolve plaque buildup, helping to prevent blockages and reduce the risk of heart attacks and strokes. Furthermore, xenobots could target the precise areas of tissue that need regeneration following a heart attack, delivering growth factors and stem cell treatments to aid recovery.

4. Infectious Diseases and Immune Therapies

Another promising area for xenobot application is in the treatment of localized infections. Xenobots could carry antibiotics or antiviral agents directly to infection sites, minimizing the risks associated with systemic antibiotic treatment, such as antibiotic resistance and disruption of gut microbiota.

In autoimmune diseases, xenobots could deliver immunosuppressive agents precisely to affected tissues, reducing systemic immune suppression and the associated risk of infection.

The Path Forward: Challenges and Ethical Considerations

While xenobots hold great promise, there are also significant challenges to address. Since they are living organisms, xenobots are susceptible to environmental factors and may require complex control mechanisms.

Ensuring the safe degradation of xenobots after completing their task is essential to avoid unintended side effects or long-term accumulation in the body. Additionally, further research is needed to understand how to precisely control xenobot behavior within the body, particularly over extended periods.

Ethical considerations are also important, as the use of living, programmable organisms within human patients is a novel and potentially controversial approach. Developing a regulatory framework to govern xenobot-based therapies will be crucial to their acceptance and successful integration into medical practice.

A Paradigm Shift in Medicine

Xenobots are at the forefront of a new era in medicine, where the convergence of biology and technology opens up possibilities once thought to be the realm of science fiction. By offering a biodegradable, programmable, and adaptable solution for drug delivery, xenobots could improve the precision and efficacy of treatments across various disease areas. As research progresses, xenobots may well become integral to personalized medicine, transforming how we treat complex and challenging conditions.

The journey of xenobots in drug delivery and targeted therapy is only beginning, but the potential they bring suggests a future where medicine is more effective, targeted, and responsive to individual patient needs than ever before.

Xenobots in Minimally Invasive Procedures

In the world of medicine, one of the most challenging aspects of surgery is making procedures as minimally invasive as possible. Minimally invasive surgery (MIS) is a method that uses small incisions, specialized tools, and innovative techniques to treat patients, aiming to reduce trauma, minimize scarring, and shorten recovery times.

This is where xenobots come in—an exciting new class of microscopic, programmable living organisms that have the potential to transform how we approach surgical procedures.

Xenobots are not robots made from metal and silicon, nor are they simply living cells. Instead, they are engineered from frog (Xenopus laevis) stem cells, carefully crafted to perform specific functions based on the way scientists arrange the cells.

By designing the structure of these cell clusters, researchers can effectively create microscopic "biobots" that swim, move, push particles, and even "self-heal" in a way that is far gentler than traditional surgical tools. With their unique design, xenobots promise to open a new frontier in the field of minimally invasive surgeries.

How Xenobots Operate in Minimally Invasive Procedures

In surgical applications, xenobots could be designed to navigate within the body, enabling them to reach hard-to-access areas. Think of xenobots as tiny, intelligent cells that can be precisely guided to the area needing treatment. Because they are composed of living cells, xenobots naturally blend with the body's biochemistry. This seamless integration enables them to navigate the bloodstream, tissues, or organs with minimal resistance, making it possible to conduct procedures without incisions or physical trauma.

Imagine a patient with a tiny tumor deep in the brain or in the center of a lung. A xenobot, capable of detecting chemical markers associated with the tumor, could navigate through the bloodstream, track down the target, and either deliver a therapeutic payload or help to break down the cells in a controlled manner.

In other cases, xenobots could be programmed to clean blocked blood vessels, deliver medicine to specific locations, or even help repair tissue micro-damage without causing collateral harm to surrounding tissues.

Advantages of Xenobots in Minimally Invasive Surgery

1. Precision Targeting: Due to their small size, xenobots can reach microscopic and complex parts of the body that are inaccessible to standard surgical instruments. This makes them an ideal solution for targeting diseased cells, such as cancer cells, at an early stage before they can grow or spread further. Their precision could transform cancer treatment, allowing targeted therapies that avoid healthy cells.

2. Reduced Infection Risk: One of the primary complications of traditional surgery is the risk of infection, which can stem from open incisions and foreign objects introduced into the body. Since xenobots are derived from living cells and are biodegradable, they do not introduce foreign material into the body.

This significantly reduces infection risks and does away with the need to remove them from the body, as they naturally break down after completing their task.

3. Natural Biodegradability: Traditional implants or surgical tools must either be removed post-operation or left inside the body, which can sometimes lead to complications.

In contrast, xenobots are biodegradable; they are designed to self-decompose after performing their function. This reduces both the long-term risks and the need for further surgical intervention to remove them.

4. Self-Healing Abilities: Unlike conventional robots, which would require costly repairs if damaged, xenobots can "self-heal" to a degree, allowing them to continue their function even if they encounter minor damage.

This property adds reliability to their use, especially in complex surgeries where human intervention may be limited or where robots face challenging micro-environments within the body.

Applications of Xenobots in Various Types of Minimally Invasive Surgeries

1. Cancer Treatment: In treating cancer, xenobots could offer revolutionary benefits. By injecting xenobots into the patient's body, they can be directed towards tumor sites to deliver chemotherapy drugs precisely, minimizing damage to surrounding healthy cells.

In future cancer surgeries, a swarm of xenobots could potentially identify, break down, or even encapsulate cancer cells, marking a significant advancement in targeted therapies.

2. Cardiovascular Procedures: Cardiovascular diseases are often treated with complex surgeries that require clearing blockages in arteries. Xenobots can be programmed to navigate through the bloodstream, identifying areas of plaque build-up or other obstructions.

Once there, they could help break down the blockage or deliver anti-coagulant medications directly to the site, avoiding the need for risky procedures like angioplasty or stent placement.

3. Neurological Interventions: The brain is one of the most delicate organs, where even minor damage during surgery can lead to significant consequences.

Xenobots offer a safer alternative, capable of delivering drugs or performing repairs within the brain's complex structure.

They could also play a role in treating neurodegenerative diseases by delivering treatments directly to affected brain cells, bypassing the blood-brain barrier that typically prevents most drugs from reaching these areas.

4. Organ Repair and Tissue Regeneration: Xenobots may aid in repairing internal organs and regenerating tissue. For instance, following a heart attack, xenobots could be guided to deliver stem cells to the heart tissue to promote recovery and help with scar tissue repair.

In other cases, they might facilitate healing after trauma by delivering growth factors or aiding tissue integration, offering new hope for patients with damaged organs.

Challenges and Future Prospects of Xenobots in Surgery

Despite their potential, xenobots face challenges before they can become mainstream in minimally invasive surgery. One of the main hurdles is ensuring their precision and safety in real-time conditions inside the human body.

Scientists are still working to perfect the control mechanisms to prevent xenobots from causing unintended side effects or veering off course.

Additionally, ethical concerns are raised regarding the use of living cells for medical procedures.

Regulatory approval will be essential, involving rigorous testing to verify that xenobots pose no long-term risks to patients.

Researchers must also address issues of scalability, ensuring that xenobot production can be achieved cost-effectively for widespread medical use.

The Vision Forward: A Minimally Invasive Future

As scientists continue to refine xenobot technology, the possibilities for minimally invasive surgeries are expanding.

The notion of complex, high-risk surgeries being performed with no incisions, minimal pain, and shorter recovery times could soon become a reality.

With further advancements, xenobots may become the preferred choice for a range of treatments, making surgery a far less invasive experience for millions worldwide.

Xenobots represent a synthesis of biological intelligence and robotic functionality that goes beyond the capabilities of any traditional tool.

With their unique blend of biological compatibility and sophisticated functionality, xenobots could very well become one of the most significant advancements in surgical care, transforming patient recovery times, enhancing precision, and fundamentally reshaping how we view surgical intervention.

Chapter 11: Ethical and Moral Considerations

The development of "xenobots" — living, programmable organisms made from frog cells — has sparked fascination, curiosity, and significant ethical discussions. These microscopic, self-powered robots are created by reassembling living cells into novel forms, enabling them to perform simple tasks, move autonomously, and even self-replicate in some cases.

While xenobots hold the potential for revolutionary applications, such as environmental cleanup, medical therapies, and understanding the fundamentals of life, their development has raised a range of ethical and moral concerns.

These concerns include the nature of life, responsibility for biological technologies, and the possible societal consequences.

The following exploration addresses these key ethical and moral questions, shedding light on what it means to create life and how humanity might approach this breakthrough responsibly.

1. The Nature of Xenobots

One of the first questions raised by xenobot technology is what defines "life." Xenobots are made entirely from living cells — typically from frog embryos — yet they do not possess a central nervous system or consciousness.

They are alive in the sense that they are made of living material, can reproduce under specific conditions, and respond to their environment, but they lack the essential traits associated with sentient beings, such as self-awareness and the ability to feel pain.

This blurs the line between machine and organism, creating a "gray area" for defining life.

For some, the idea of creating a biological entity without a clear understanding of its moral or ethical status is unsettling. Is it ethical to design and manipulate life forms solely to serve human purposes, even if these life forms lack consciousness? This raises the broader question of whether life — in any form — holds intrinsic value, or if value is derived only from an entity's sentience, autonomy, and capacity for suffering.

2. The Limits of Human Intervention: "Playing God"?

The creation of xenobots also confronts humanity with an age-old ethical concern: are we "playing God" by creating new life forms? This question often arises when scientific advances push the boundaries of what was previously thought possible. In the case of xenobots, scientists have not only manipulated but also redesigned cells into entirely new forms that would never naturally occur.

Some people argue that there should be limits to human intervention in nature. The fear is that tampering with the fundamental building blocks of life may have unpredictable, unintended consequences, both for the xenobots themselves and for the ecosystems they could affect. Others counter that scientific inquiry and technological development are inherently human activities, and that responsible experimentation with life forms, as long as it is carefully controlled, is a natural extension of scientific progress. However, this argument raises further questions: how do we determine what is "responsible," and who has the authority to decide these limits?

3. Potential for Misuse: Dual-Use Technology

As with many advanced technologies, xenobots have potential applications that can be either beneficial or harmful. In the hands of well-intentioned scientists and policymakers, xenobots could aid in cleaning up microplastics from oceans, delivering drugs precisely within the human body, or even repairing damaged tissues. However, the technology could also be misused, especially in contexts where control over living, self-replicating organisms might lead to unintended consequences.

Dual-use technologies, those that can be used for both good and harm, present unique ethical dilemmas. Xenobots' potential for self-replication, for example, raises concerns about ecological risks and the possibility of xenobots escaping human control. Even though they are currently unable to survive outside laboratory conditions, future advancements could change this, making containment a crucial ethical consideration. Establishing strict regulatory frameworks and enforcing transparent oversight are vital steps to prevent misuse and reduce the risks associated with xenobot deployment.

4. Responsibility Toward Newly Created Life Forms

If we create living, autonomous organisms, do we owe them any form of moral consideration or responsibility? This question lies at the heart of the xenobot debate, as it requires rethinking traditional ideas of responsibility and compassion. Unlike sentient animals or even artificial intelligences with complex programming, xenobots lack a nervous system, pain receptors, or any capability for consciousness. By traditional ethical standards, this would imply that they do not need protection or consideration.

Yet, some ethicists argue that, as creators of xenobots, we may have a duty to ensure that we do not harm them without cause, even if they are unaware of their own existence. This perspective challenges us to consider whether the mere act of creating life confers a responsibility on the creator, irrespective of the created being's awareness or capacities. In this light, we might need to extend ethical guidelines for xenobot research, just as we have for animal testing and human trials, ensuring that any intervention is truly necessary and justified.

5. The "Slippery Slope" Argument: Where Do We Draw the Line?

A common ethical argument in the xenobot discourse is the "slippery slope" concern. If creating living, programmable organisms is permissible now, how far will we go in the future? Could this lead to more complex biological creations, potentially with forms of proto-consciousness, or even modified humans? The slippery slope argument suggests that society needs to be vigilant about setting boundaries today to prevent ethically questionable practices in the future.

While the slippery slope argument is speculative, it underscores the importance of cautious, transparent progress. Xenobot research needs clear ethical guidelines that take into account both current and potential future advancements. Scientific inquiry, while valuable, must be tempered by ethical foresight and a commitment to avoid unintended consequences.

6. Public Engagement and Trust

Another critical ethical consideration involves public transparency and engagement. The development of xenobots, like any transformative technology, has the potential to affect society in profound ways. It is essential for scientists, ethicists, and policymakers to engage with the public to explain what xenobots are, why they are being developed, and how they will be regulated.

Public engagement not only builds trust but also ensures that society as a whole has a say in how such technology is deployed. Involving diverse perspectives — including environmental advocates, ethicists, religious leaders, and laypeople — can help ensure that xenobot technology is used responsibly and in a manner that aligns with societal values.

Balancing Innovation and Responsibility

The ethical and moral questions surrounding xenobot research are complex and multifaceted. These tiny, living robots challenge us to consider what it means to create life, our responsibilities as creators, and the limits of human intervention in natural processes. Xenobots present enormous potential benefits, but they also carry risks that cannot be overlooked. Thoughtful, ethical consideration is necessary to ensure that we navigate these issues responsibly.

As xenobot research progresses, ongoing dialogue, public engagement, and the establishment of clear ethical frameworks will be critical. This balanced approach can allow society to harness the potential of xenobots while safeguarding against risks and respecting the values and principles that govern human progress. In navigating these questions thoughtfully, we can help ensure that the age of xenobots unfolds with care, foresight, and respect for life.

Exploring the Ethical Implications of Creating Living Machines

In recent years, a breakthrough in biotechnology has emerged: xenobots, tiny biological robots made from frog cells, carefully designed to move, sense, and even self-replicate under certain conditions. This discovery has raised exciting possibilities for medical, environmental, and technological advancements.

But alongside this excitement comes a series of profound ethical questions. As we venture into a future where biological machines might play a role in human life, it's crucial to examine the potential risks and moral implications of creating living machines.

1. Defining Life: Where Do Xenobots Fit?

At the heart of the ethical debate is the question, "What does it mean to be alive?" Xenobots, created from the cells of African clawed frogs, aren't organisms in the traditional sense. They don't have brains, complex organs, or the ability to grow independently outside of lab environments. However, they do exhibit characteristics we usually associate with life, like movement, behavior in response to stimuli, and even a basic form of reproduction.

While they lack self-awareness, they blur the boundary between machines and living organisms. This ambiguity raises ethical concerns around the responsibility we hold in creating something that behaves in ways traditionally exclusive to living beings. Is it morally acceptable to use life for our technological aims?

2. Biological Machines in Medicine and the Environment

Xenobots could offer remarkable benefits. In medicine, they have the potential to deliver drugs to specific sites in the body, perform microsurgery, or even repair damaged tissues. In the environment, they could help in cleaning up microplastics from oceans or gathering pollutants. These applications present potential for significant good, but there's an ethical catch. Deploying biological machines in human bodies or natural ecosystems involves a level of unpredictability.

Once xenobots are released into a biological system, there may be risks of unintended interactions. For example, what if they mutate or evolve in unforeseen ways? Could they disrupt ecosystems or harm human cells in unforeseen ways?

The principle of "first, do no harm" is especially relevant here, urging scientists to prioritize safety, control, and reversibility when developing xenobot technology.

3. The Right to Create and Terminate Living Machines

The idea of creating life—however minimal or mechanical—introduces difficult questions about moral rights. Do xenobots have rights, however limited? Although xenobots are currently very simple biological systems with no nervous system or consciousness, as this technology advances, it may one day become possible to engineer biological machines with more complex functions, potentially resembling simple organisms with some capacity for sensory experience.

Does creating a self-replicating, mobile xenobot give it a claim to life? If not, what threshold would need to be crossed before such a claim could be made? While xenobots are currently "terminated" once they are no longer useful, more complex biological machines might provoke stronger ethical reactions.

This consideration leads to a broader question about humanity's right to terminate entities it creates, especially if those entities possess life-like qualities.

4. Playing God: Are Humans Overstepping Boundaries?

A recurring theme in the ethics of biotechnology is the question of whether scientists are "playing God." Critics of xenobot technology might argue that creating life in a laboratory setting shows a lack of humility toward nature and may have far-reaching consequences that we cannot yet foresee.

By designing and programming living forms to carry out our commands, are we imposing human will over life in a way that should be restricted?

This concern isn't simply about religion or philosophy—it's a practical question about respect for life and the dangers of overestimating human control over biological systems.

Biotechnology often comes with unintended consequences, and in the case of xenobots, the ability to create and modify life at a cellular level brings new dimensions of responsibility and potential for error.

5. The Slippery Slope: Toward a Future of Bioengineering?

The xenobot project represents only one step toward the future of bioengineering. Today's xenobots are simple structures, but tomorrow's biobots may be much more complex, potentially capable of self-regulation or responding to environmental cues in sophisticated ways.

As scientists develop the ability to create increasingly complex biological machines, society may find itself on a "slippery slope" toward a world where engineered life forms become common. This might lead to ethical dilemmas around the preservation of "natural" life versus "engineered" life and whether the two should be regarded differently.

Some argue that xenobot technology could ultimately shift our relationship with life itself. If we start to see life as something that can be engineered to serve specific functions, will this lead to a devaluation of natural organisms? And if living machines become prevalent, could they one day replace natural ecosystems or traditional animal roles, disrupting ecological balances?

6. Regulatory and Ethical Safeguards: A Necessary Foundation

To address these concerns, it's essential to establish ethical guidelines and regulatory frameworks. Several fundamental questions need answering: Who gets to decide what kinds of living machines are acceptable to create?

Should there be limits on how these machines are used or on the complexity they can achieve? Establishing these rules early on can help prevent misuse and set boundaries for responsible innovation.

One proposed approach is to apply "precautionary principles" in xenobot research. Scientists could be required to provide thorough risk assessments and proof of environmental and health safety before deploying xenobots in real-world applications.

Additionally, transparency around research findings, public engagement, and interdisciplinary input from ethicists, biologists, and engineers can help ensure that xenobot technology is developed responsibly.

7. A Hopeful Vision, Grounded in Responsibility

Despite these ethical challenges, the potential of xenobots is immense. They could be the key to innovations in medicine, ecology, and technology that benefit humanity and the planet. With careful oversight, xenobots might one day help us tackle issues like pollution, resource scarcity, and medical limitations in ways that are safe and sustainable.

Ethical science requires a balance between innovation and humility, an awareness that new technologies should serve life rather than dominate it. For xenobot researchers, this means embracing the responsibility that comes with creating living machines.

By respecting the limits of their knowledge and the unpredictable nature of biological systems, scientists can work to minimize risks and prioritize outcomes that enhance, rather than threaten, life on Earth.

As we step into this new era of "living machines," we are faced with complex questions that science alone cannot answer. The ethical implications of xenobot technology compel us to rethink our relationship with life, agency, and the natural world.

By approaching these developments with caution, transparency, and a commitment to ethical principles, we can explore the potential of xenobots without losing sight of the deeper responsibilities they carry.

It's up to humanity to decide how far we go in this journey—and to ensure that our pursuit of innovation aligns with values that respect and preserve life.

Balancing Benefits and Risks in Xenobot Research

Xenobots, tiny living robots created from the stem cells of the African clawed frog (Xenopus laevis), represent a groundbreaking advancement in biotechnology and synthetic biology. As remarkable as they are, xenobots also pose complex ethical, biological, and environmental questions.

Striking a balance between their potential benefits and the risks they introduce is essential to ensure their responsible development and deployment.

In understanding this balance, it's important to explore the science behind xenobots, their potential applications, and the specific concerns that have arisen in response to this innovative field of research.

The Science Behind Xenobots

Xenobots are not mechanical robots in the traditional sense; they are made entirely of living cells, structured and programmed to behave in certain ways. These micro-organisms, typically less than a millimeter wide, are constructed from frog stem cells that are cultured, separated, and arranged into specific forms using micro-surgical tools and processes.

With the help of computer algorithms, scientists can design xenobots to complete specific tasks based on their shape, composition, and cell type. Xenobots can move autonomously, carry microscopic payloads, and even self-repair minor damage, making them highly versatile.

The regenerative nature of xenobots also contributes to their uniqueness. Unlike traditional robotic devices, they can adapt, fuse, and even "heal" if damaged. These properties open the door to using xenobots as tools for various applications, especially those requiring microscopic precision.

Their degradable biological composition means they can eventually disintegrate, leaving no toxic materials behind—an attribute that aligns with eco-friendly, sustainable practices.

Potential Benefits of Xenobot Research

The potential benefits of xenobot research are substantial and cover several areas:

1. **Medical Applications:** Xenobots could revolutionize targeted drug delivery and surgery. They can be programmed to deliver drugs to specific areas within the human body, potentially improving treatment efficacy while minimizing side effects.

Additionally, xenobots may one day assist in microsurgeries, navigating through the bloodstream or tissue to repair damage or remove clots. Their self-healing property could reduce the need for additional interventions during surgeries.

2. **Environmental Applications:** Xenobots might serve as living machines to address pollution. Since they can move autonomously and target specific particles, xenobots could help clean up microplastics in water bodies.

Their biodegradable nature makes them an eco-friendly alternative to synthetic robots or chemicals in pollution remediation.

3. **Scientific Research and Development:** Xenobot research helps scientists better understand cellular behavior, morphogenesis, and tissue engineering.

The knowledge gained could lead to advances in regenerative medicine and provide insights into developmental biology. Xenobots also serve as a model to study self-organizing systems, potentially inspiring new developments in artificial intelligence.

4. **Evolution of Robotics:** Xenobots introduce an entirely new realm of robotics that operates within biological and ecological boundaries, allowing scientists to think beyond traditional hardware and software. By blending biological tissues with artificial intelligence, researchers can explore the future of biohybrid technologies, possibly leading to advanced robotic systems that can self-repair and adapt to environmental changes.

Risks and Ethical Concerns in Xenobot Research

Despite the promising benefits, xenobot research is not without risks and ethical concerns. These concerns revolve around biosafety, bioethics, environmental impact, and unintended consequences.

1. Unintended Evolution and Mutations: Although xenobots are designed for specific tasks, there is a risk of unintended mutations or adaptations, particularly if released into uncontrolled environments. Evolutionary pressures might cause xenobots to change their behavior or physical structure in unexpected ways, potentially leading to harmful interactions with natural ecosystems.

Scientists need to carefully monitor xenobot behaviors to prevent unanticipated adaptations that could disrupt ecological balances.

2. Containment and Environmental Risks: While xenobots are biodegradable, the possibility of environmental contamination cannot be ignored. If xenobots were unintentionally released into ecosystems, they could interact with native species, possibly altering microbial communities, nutrient cycles, or even the health of specific organisms. It's essential to assess the ecological impacts thoroughly before considering any environmental applications.

3. Bioethical Implications: Xenobots blur the line between living and synthetic entities, raising ethical questions about the nature of life, autonomy, and control. Are xenobots "alive," or are they simply tools? If they exhibit autonomous behavior, do they have intrinsic value or rights? These questions are largely philosophical but have practical implications in setting regulatory standards and guidelines for xenobot research.

4. Security Risks and Bioterrorism: Xenobots could potentially be repurposed for harmful applications, such as carrying toxic substances or spreading pathogens. If xenobot technology falls into the wrong hands, it could be misused to harm individuals or disrupt ecosystems.

Ensuring that xenobot research and development are closely monitored and regulated is crucial to prevent potential misuse.

5. **Human Health and Safety:** Since xenobots are composed of biological cells, there is always a small risk of immune reactions or other adverse effects if they are used in human applications. While most xenobot designs are unlikely to provoke strong immune responses, rigorous safety testing is essential before any medical use.

Approaches to Risk Mitigation and Ethical Oversight

To responsibly advance xenobot research, scientists and policymakers are exploring various strategies to address these risks and ensure ethical standards are met.

1. **Strict Containment Protocols:** Researchers are developing containment measures for xenobots, especially during testing phases. Lab-based containment protocols help ensure that xenobots do not escape into unintended environments.

For applications involving environmental use, scientists are focusing on creating xenobots that degrade more quickly and lose functionality outside controlled settings.

2. **Ethical Review Boards and Oversight:** Many institutions conducting xenobot research have established ethical review boards to address the moral dimensions of this work. These boards typically include bioethicists, scientists, and public representatives who assess the potential societal impacts and guide the research accordingly. Involving diverse perspectives helps ensure a balanced approach to both the scientific possibilities and ethical considerations.

3. **Biosafety Regulations and International Collaboration:** Establishing international standards and guidelines is crucial to governing xenobot research. Organizations like the World Health Organization (WHO) and the United Nations (UN) could play roles in regulating the development and distribution of xenobot technology to prevent misuse and unintended consequences.

Collaboration among countries can create a more comprehensive approach to xenobot research, with guidelines that reflect a global consensus on safety and ethics.

4. Public Engagement and Transparency: Public education and transparency in xenobot research are essential to address societal concerns. Clear communication about the goals, methods, and safety measures involved in xenobot research can build public trust and alleviate fears.

Providing opportunities for public input, such as through community forums, can also foster a more inclusive approach to research.

Xenobot research sits at a fascinating but complex intersection of biology, robotics, and ethics. While xenobots offer transformative potential in fields ranging from medicine to environmental science, they also raise significant ethical and practical challenges.

The future of xenobot research will depend on a careful balance between the desire for innovation and the need for responsible oversight. By developing stringent guidelines, engaging the public, and fostering a global dialogue, scientists and policymakers can work together to realize the benefits of xenobots while minimizing the associated risks.

Chapter 12: Xenobots and the Future of Biotechnology

Imagine a future where small, programmable organisms serve as living tools for health, environmental cleanup, and more. This isn't science fiction but a glimpse into the transformative field of biotechnology, driven by a remarkable innovation: xenobots. Xenobots, named after the African clawed frog Xenopus laevis, are tiny, self-organizing biological robots created from frog cells. They are not robots in the traditional sense but are composed of living cells that move, heal, and work together under programmed instructions, like biological machines. First developed in 2020 by a team of scientists from Tufts University and the University of Vermont, xenobots represent an extraordinary fusion of cellular biology, artificial intelligence, and robotics. They hint at a new era in biotechnology, where living systems can be shaped to solve some of humanity's most pressing challenges.

The Biology Behind Xenobots

Xenobots are built from two primary cell types: skin cells and cardiac cells of frog embryos. Skin cells provide structural support, while cardiac cells, which contract rhythmically, act as tiny engines that drive the xenobots' movement. Researchers start with a collection of these cells, then use computer simulations to design structures that can achieve specific goals, such as walking, swimming, or carrying small particles. After a virtual model is developed, scientists assemble the cells in a lab to form xenobots that mimic the simulation's design.

Because xenobots are made from living cells, they can perform tasks that traditional robots struggle with, like self-repair. If a xenobot is cut in half, it can heal itself, potentially extending its life and utility. Additionally, xenobots are biodegradable, breaking down harmlessly over time, unlike conventional materials that persist in the environment. This natural decomposition makes them highly attractive for ecological applications, as they can accomplish tasks without leaving a toxic footprint.

How Artificial Intelligence Shapes Xenobot Design

The creation of xenobots relies heavily on artificial intelligence (AI). Using evolutionary algorithms, a type of AI, researchers create thousands of potential designs, each with variations that may impact the xenobots' functionality.

These designs are tested in virtual environments to find the best forms that could, for example, move efficiently or transport microscopic payloads. The AI-based simulations allow scientists to explore a range of shapes and movement types that might be too complex or time-consuming to build and test manually.

The AI doesn't just identify successful designs; it essentially "evolves" xenobots by selecting and refining the best-performing ones across multiple generations.

This computational process speeds up what would otherwise be a trial-and-error effort, enabling scientists to create highly specialized xenobots that are tailor-made for specific tasks.

The combination of biology and AI results in living machines that are adaptive, efficient, and more effective at performing certain tasks than traditional robots.

Applications of Xenobots in Medicine

One of the most exciting potential uses for xenobots lies in the medical field. Because they are made from biological material, xenobots could, in theory, perform medical tasks within the human body without triggering immune responses, unlike synthetic materials.

Researchers are exploring ways to use xenobots for drug delivery, where they could transport medicine to precise locations within the body, minimizing side effects and increasing treatment effectiveness.

They might also be used to clear out harmful plaque in arteries, potentially reducing the risk of heart attacks, or target and destroy cancer cells in a highly targeted manner.

Another promising avenue is wound healing. Xenobots could be programmed to move towards injury sites and release compounds that accelerate the healing process.

Their ability to self-repair could allow them to survive longer within the body, offering prolonged therapeutic effects without needing to be replaced.

This feature also makes xenobots potentially useful for non-invasive surgeries or procedures where tiny agents could perform complex tasks within the body without cutting or stitching.

Environmental Benefits and Potential Applications

Beyond medicine, xenobots offer remarkable opportunities in environmental biotechnology. Because they are biodegradable and non-toxic, they could be deployed in ecosystems without long-term ecological impacts.

Xenobots could be used to clean up microplastics in oceans or other waterways by moving through water and collecting or breaking down plastic particles.

Traditional cleanup methods are costly, slow, and often only remove large pieces of plastic, but xenobots, as biological agents, could work continuously and at a microscopic level.

Another ecological application could be xenobot-assisted bio-remediation, where xenobots locate and break down pollutants, from oil spills to hazardous chemicals in water and soil.

The xenobots' adaptability and precision make them ideal for working in fragile ecosystems where other cleanup methods might cause damage.

In theory, researchers could even program xenobots to detect and neutralize harmful microorganisms, reducing the spread of infectious diseases in both human and animal populations.

Ethical Considerations and the Path Forward

As with many biotechnological advancements, xenobots raise ethical questions that need careful consideration. Since xenobots are living entities, questions arise regarding their status and use.

Are they tools, organisms, or something in between? Should they be subject to ethical guidelines similar to those governing research on animals?

Additionally, the prospect of creating self-replicating xenobots brings potential risks. While xenobots cannot reproduce independently at present, future developments may lead to versions that can.

If not carefully controlled, self-replicating xenobots could disrupt ecosystems or outcompete natural organisms, leading to unintended consequences.

Regulation is crucial to ensure that xenobot technology develops safely and responsibly. Many scientists advocate for transparent research practices, strict oversight, and global guidelines to prevent misuse.

These frameworks would aim to harness xenobots' potential while minimizing risks, ensuring that they are used for beneficial purposes and not harmful applications.

The Future of Xenobots in Biotechnology

As xenobot research advances, their applications and design will likely become increasingly sophisticated.

In the near future, we could see xenobots with sensory capabilities that respond to stimuli, such as light or chemical signals, making them even more precise in complex environments.

Xenobots might also be equipped with genetic modifications, allowing them to perform advanced biochemical tasks or even serve as "living factories" for producing drugs, enzymes, or other beneficial compounds.

While still in its early stages, xenobot technology demonstrates the potential of combining living cells with computer science to solve both medical and environmental problems. As we continue to refine their capabilities, xenobots may become valuable, versatile tools that help address the challenges of tomorrow.

In doing so, they promise to reshape biotechnology, not just through mechanical innovation but through the sustainable and intelligent use of life itself. The journey of xenobots is just beginning, but they offer a hopeful glimpse of a future where biology and technology work seamlessly together to build a better world.

Speculating on the Long-Term Impact of Xenobots on Biotechnology

The potential of xenobots, biological robots made from living cells, has captivated scientists and the public alike. Born from the cells of the African clawed frog (Xenopus laevis), these tiny machines have redefined the boundaries of synthetic biology and robotics.

Xenobots are programmable, self-assembling, and biodegradable, providing a glimpse into a future where biological machines might revolutionize fields as diverse as medicine, environmental science, and synthetic biology.

Here, we'll explore the long-term implications of xenobots for biotechnology and the profound shifts they could bring.

1. Revolutionizing Drug Delivery and Precision Medicine

Xenobots could transform drug delivery by providing a highly controllable, targeted system for delivering medications. Unlike synthetic drug carriers, xenobots are entirely organic, meaning they could degrade naturally in the body without causing harm.

Imagine a scenario where xenobots, programmed to recognize cancer cells, transport chemotherapy drugs directly to a tumor. This would drastically reduce the systemic side effects seen in traditional chemotherapy, enhancing patient well-being.

Moreover, xenobots could potentially be adapted to monitor a patient's internal environment in real-time, responding dynamically to biochemical changes.

This adaptability may enable more precise, personalized treatment plans tailored to each individual's biology, moving medicine closer to the goal of precision medicine.

Xenobots could become an integral part of the human body, acting as both scouts and defenders, locating disease and delivering treatments with an intelligence not yet possible in synthetic drugs.

2. Advancements in Regenerative Medicine and Tissue Engineering

Xenobots' ability to self-assemble and repair themselves points to breakthroughs in regenerative medicine. The core of regenerative medicine involves restoring lost or damaged tissues, a task xenobots may be particularly suited for.

Researchers could use xenobot technology to direct cell growth in damaged tissues, serving as scaffolding that encourages natural regeneration.

For instance, xenobots might help bridge gaps in nerve or muscle tissue, potentially aiding in recovery from severe injuries that the human body struggles to repair on its own.

Xenobot technology could also enable novel forms of bioengineering. If scientists find ways to control the differentiation of xenobots' cells, they could create custom biological structures within the body, like veins or other connective tissues.

The implications for conditions that cause structural damage in the body, such as heart disease, spinal injuries, or even organ failure, are profound. Xenobots, acting as adaptable cellular scaffolding, could facilitate tissue recovery and foster entirely new treatments in the realm of regenerative medicine.

3. Environmental Impact and Ecological Restoration

Xenobots offer unique solutions to environmental challenges, as they could be deployed to clean up ecosystems without contributing to pollution themselves.

Being biodegradable, xenobots would naturally degrade after their tasks, avoiding the issue of microplastic or toxic waste left by traditional cleanup methods.

Xenobots could be programmed to consume specific pollutants, like oil or plastics, and break them down into less harmful substances. These microscopic biological machines could clear waterways or even travel through contaminated soils, absorbing heavy metals or toxic compounds.

Further, xenobots may facilitate ecological restoration efforts by acting as mobile vehicles for carrying beneficial bacteria, seeds, or nutrients to distressed ecosystems.

For instance, they could carry and plant seeds in areas affected by deforestation or carry beneficial bacteria to degraded soils, promoting ecological health and biodiversity.

In this way, xenobots could help humanity repair the environmental damage caused by industrialization and urbanization, transforming the landscape of environmental biotechnology.

4. Shaping the Future of Synthetic Biology and Bio-Computing

Xenobots represent an intriguing bridge between biological life and digital technology, an area of biotechnology known as bio-computing.

They demonstrate that biological systems can perform tasks typically associated with machines or computers, like navigation and autonomous decision-making.

This opens up new possibilities in synthetic biology, where organisms or tissues might perform complex computational functions, bringing together biology and computer science.

One exciting area of research is bio-computing using xenobots to gather and process data directly in biological systems.

Imagine xenobots that measure environmental factors like temperature or pH within the body, storing and transmitting this data to health-monitoring devices.

Scientists are beginning to see how xenobots could act as bio-sensors, providing insights into health and environmental conditions in ways digital technology cannot.

This technology could create a world where medical diagnostics and ecological monitoring are seamlessly embedded into living systems.

5. Ethical Considerations and Potential Risks

The rise of xenobot technology presents ethical and social considerations. As xenobots become more autonomous and capable, questions about control and unintended consequences emerge.

Could xenobots evolve behaviors not originally intended by their creators? Might they interact with natural organisms in unexpected ways, causing harm to ecosystems or public health?

These concerns require a framework for regulating xenobot development and deployment, ensuring that innovation does not outpace ethical and safety considerations.

Moreover, as xenobots are created using living cells, there are philosophical questions regarding the nature of life. Xenobots blur the line between machine and organism, raising questions about whether they should be considered "alive."

This has implications for how they are treated and used, especially in sensitive applications such as medical treatments.

Balancing progress with ethical caution will be crucial to ensuring that xenobots fulfill their potential while respecting biological integrity.

6. Educational and Societal Impact

The development of xenobots is already inspiring the next generation of scientists, and as their potential becomes clearer, they are likely to fuel public fascination with biotechnology. Xenobots provide a tangible example of synthetic biology's possibilities, making complex science accessible to students and the public.

As a result, xenobot technology could enhance scientific literacy and inspire young people to pursue careers in STEM fields, accelerating progress in biology and medicine.

At a societal level, the introduction of xenobots will inevitably shift public attitudes toward biological manipulation and artificial life. Society will need to engage with the ethical and philosophical implications of creating artificial organisms, fostering an informed debate about how biotechnology should be integrated into daily life.

Xenobots offer a glimpse of a future where biology and technology are inseparable, underscoring the need for policies that govern this new relationship thoughtfully and responsibly.

The long-term impact of xenobots on biotechnology is profound, touching nearly every aspect of modern life. From revolutionizing medicine and environmental science to raising new ethical questions, xenobots symbolize the dawn of a new age in synthetic biology.

They embody the potential for biology to operate as both a machine and a natural system, bridging gaps between technology and life in ways previously thought impossible.

While the journey toward integrating xenobots into society will undoubtedly bring challenges, their promise points to a future where biotechnology not only enhances our health and environment but redefines our relationship with the living world.

The path forward is one of innovation, responsibility, and wonder, as we embrace this powerful new frontier in biotechnology.

Potential Challenges and Breakthroughs in the Field

The creation of xenobots—a fusion of biology and robotics—has sparked fascination and promises profound applications across fields from regenerative medicine to environmental repair.

However, advancing this field comes with significant scientific, ethical, and technical challenges. These challenges often act as barriers to rapid progress, yet each challenge also opens the door to potential breakthroughs that could transform how we understand and utilize living systems.

Below is a closer look at some of the pressing obstacles and the transformative potential that xenobots may unlock.

1. Technical and Design Challenges

One of the primary technical challenges in xenobot research is creating predictable and reliable designs that can function autonomously in complex environments. Xenobots, composed of living cells from frog embryos, have limited lifespans, and their behavior in various environments is challenging to control. Constructing a "programmed" organism that can consistently perform complex tasks requires deeper understanding and manipulation of cellular biology.

To address this, researchers are exploring computational modeling and AI-driven design. By using algorithms to simulate xenobot structures and behaviors, scientists can experiment with different configurations virtually before implementing them in a lab. Such models offer a powerful tool to minimize trial and error and enable more targeted designs.

This combination of biology and computational power has already led to breakthroughs in design predictability, with some xenobots now able to demonstrate controlled movement and specific task-oriented behaviors. Further advancements in modeling could yield xenobots capable of performing more precise and customizable actions, from tissue repair to targeted drug delivery.

2. Longevity and Sustained Function

Xenobots currently have limited lifespans, often surviving only days or weeks, which restricts their practical applications. This poses a challenge for tasks that require sustained activity over longer durations, such as environmental clean-up or persistent medical interventions. Researchers are experimenting with cell types, exploring genetic modifications, and testing cell combinations to extend xenobot lifespans. One breakthrough approach is the use of cells with regenerative properties, such as stem cells, which might allow xenobots to self-repair or maintain their functionality over extended periods.

While lengthening xenobot lifespans is a challenge, achieving it could open a range of applications previously considered out of reach. Extended lifespans could enable xenobots to serve as long-term bio-devices for continuous monitoring within the human body or ecosystems.

Sustained function could also make xenobots viable in large-scale environmental interventions, such as collecting microplastics from oceans or aiding in the repair of damaged ecosystems.

3. Biocompatibility and Ethical Concerns

As xenobots are living organisms, there is an inherent risk of unintended consequences if they are introduced into natural or human environments. Biocompatibility is crucial, especially for medical applications where xenobots might interact directly with human tissues. The immune response they could trigger, as well as the potential for contamination or disruption of local ecosystems, are important areas of study.

To address these concerns, researchers are focusing on ensuring that xenobots are biodegradable and designed to dissolve after completing their task. Another ethical approach being pursued is the use of "kill switches," programmed mechanisms that ensure xenobots cease functioning if they stray from their intended purpose. Advances in genetic engineering also make it possible to restrict xenobots' lifespans or capabilities to prevent unintended reproduction or self-sustaining populations.

The ethical and ecological challenges surrounding xenobots are prompting wider discussions among scientists, policymakers, and the public.

A breakthrough in developing fully controllable and safe xenobots would require transparent protocols and potentially new regulatory frameworks.

Success here could pave the way for xenobots to be deployed responsibly in sensitive settings, from the human body to protected natural environments.

4. Scaling Production and Cost Reduction

Xenobot production currently relies on meticulous lab work, requiring skilled technicians and costly biological materials.

Scaling up xenobot production for widespread applications, such as environmental remediation or personalized medicine, would demand more efficient, automated manufacturing processes.

One potential breakthrough in this area is the development of biofactories—systems that could mass-produce xenobots using automated assembly techniques.

Advances in tissue engineering and bio-printing technologies might make it possible to scale up production while maintaining quality and functionality.

As production becomes more efficient, the cost of xenobot manufacturing could decrease significantly, making them more accessible for various applications, from healthcare to environmental conservation.

Scaling up xenobot production would also enable researchers to perform larger studies, leading to a deeper understanding of their capabilities and limitations.

It could provide more consistent data on xenobot behavior in diverse conditions, helping to refine their designs and expand their usability.

5. Intelligence and Autonomous Function

For xenobots to function independently in complex environments, they require a certain degree of intelligence or at least programmed behavioral responses. Currently, xenobots operate based on their initial cellular organization and cannot respond dynamically to changing environments.

Developing xenobots with the ability to sense and adapt to their surroundings remains a formidable challenge but also a promising area for breakthroughs.

Researchers are experimenting with ways to incorporate basic decision-making capabilities within xenobots, such as responding to specific chemical cues or physical obstacles. The integration of simple neural-like networks—essentially, creating "brains" for xenobots—could enable them to navigate their environment with more sophistication.

This would represent a breakthrough in bio-robotics, enabling the creation of adaptive, smart xenobots that could respond to changes in real time. Such functionality could make xenobots valuable tools in unpredictable environments, such as disaster zones, where they might assist in locating survivors or delivering critical supplies.

6. Environmental and Ecological Impact

The use of xenobots in large-scale applications raises concerns about their potential impact on natural ecosystems. If released in significant numbers, even bio-degradable xenobots could disrupt local microbial populations or interfere with delicate ecological balances. Assessing these risks accurately and creating guidelines to mitigate them is essential.

Some researchers are focused on designing xenobots that are environmentally safe by using biodegradable materials and restricting their active lifespan. Advances in this area could lead to "green" xenobots that safely decompose without leaving any harmful residue.

Ensuring that xenobots contribute positively to the environment rather than introducing new risks would be a major breakthrough, fostering public acceptance and enabling more widespread use in ecological restoration projects.

The field of xenobot research is an evolving landscape of challenges and breakthroughs. The obstacles in areas like longevity, biocompatibility, and scalability are formidable, yet every breakthrough opens new possibilities.

Through interdisciplinary collaboration and ongoing innovation, researchers are steadily transforming the vision of xenobots from a novel concept into a practical tool with vast potential.

Addressing these challenges with scientific rigor and ethical foresight will be crucial to harnessing the full promise of xenobots and paving the way for their responsible integration into medicine, environmental science, and beyond.

Chapter 13: Collaborations between Biologists and Engineers

The quest to create xenobots—living, programmable organisms—the partnership between biologists and engineers has been nothing short of transformative. Biologists, with their understanding of the natural world and cellular behavior, and engineers, with their expertise in design and control systems, have joined forces to push boundaries in both biology and robotics.

This collaboration is reshaping how we think about "machines" and "organisms," blending the two into a new frontier of bioengineering.

Foundations of Bio-Engineering Partnerships

The starting point of creating xenobots involves fundamental biological research, where biologists work closely with engineers to identify which cells are best suited for programmable functions. Typically, frog cells (often from the species Xenopus laevis) are used, given their robustness and ability to be cultured outside of a traditional biological environment. By studying these cells in controlled laboratory conditions, biologists have identified key properties, such as cell adhesion, growth rates, and natural behaviors that can be manipulated for robotic functions.

Engineers then step in to create frameworks to control and shape these cells. Using principles from mechanical engineering, such as structural design and materials science, they work with biologists to construct xenobots with forms that support specific movements or functions.

For instance, engineers design cellular molds or use micro-surgical tools to craft the desired shapes, while biologists monitor cell health and functionality. This back-and-forth exchange allows each discipline to lend their expertise and learn from the other's approaches.

Computational Models: Designing in Silico

Another core area where biology and engineering intersect is in the computational phase. Before any real cells are used, engineers and computer scientists often create detailed simulations or "in silico" models to predict how different cellular configurations will behave. These models allow teams to test countless shapes, movements, and tasks without the need for physical prototypes.

With biological insights on cellular behavior, engineers build algorithms that simulate potential xenobot structures. These models use computational principles that can predict how thousands of individual cells might move and work together as a single unit. By using these predictive models, engineers can "design" a xenobot virtually and then test its predicted behavior in a simulated environment. Only after these virtual tests suggest promising results do biologists begin the process of creating physical versions of these xenobots.

Building Living Machines: Lab Synergy

Once a promising xenobot design is identified, the real work in the lab begins. Engineers and biologists come together in a shared laboratory environment, where their roles frequently overlap.

Engineers bring in tools like micro-manipulators, which help to shape clusters of cells into forms that biologists have identified as potentially functional. The tools used here are highly specialized, as the manipulation of living cells is delicate and requires precise control.

As the cells begin to take shape, biologists observe them for signs of growth, communication, and behavior. Their expertise is crucial here—biologists can recognize if cells are thriving, and if not, they can suggest adjustments to the structure or even to the chemical environment to foster cell health. Engineers, on the other hand, help modify the design as needed based on these observations, making the interaction dynamic and iterative. The resulting xenobots, typically no larger than a grain of sand, are unique in that they can heal themselves, swim, or even carry small payloads.

Programming Biological Functions

One of the most groundbreaking aspects of xenobot creation lies in programming. But rather than programming a traditional robot with code, biologists and engineers here are working to "program" cells at a biological level. This can involve manipulating how cells communicate with each other or how they react to external signals. Biologists draw from fields such as synthetic biology and molecular genetics to understand and control the mechanisms at play.

For instance, if a xenobot design requires the ability to move in response to light, the biologist-engineer team might genetically modify cells to express light-sensitive proteins. Engineers, meanwhile, craft the environmental conditions or apply specific stimuli, like changes in light, to trigger movement.

This programming doesn't involve conventional electronics but uses biological "switches" that respond predictably. This bio-programming is vital in creating xenobots that can perform simple, predictable tasks—laying the groundwork for more complex future functions.

Challenges and Problem Solving

Creating xenobots is not without its challenges. Cells are complex, and unlike traditional robotic parts, they behave unpredictably in certain environments.

For example, while engineers may design a xenobot to move in one direction, the cells may respond differently, creating erratic or inefficient motion.

Engineers must refine their designs based on these observations, but they often rely on biologists to interpret why cells may behave a certain way under specific conditions.

Similarly, as xenobots are made entirely from biological materials, they are prone to degradation or unintended interactions with other cells or environments.

Biologists play a crucial role in addressing these issues, advising engineers on modifications that could make the xenobot more resilient or able to operate in different environments.

By working together, they develop innovative solutions to extend the xenobot's lifespan and function.

Future Directions and Opportunities

The potential applications for xenobots range from environmental cleanup to medical diagnostics, and both biologists and engineers are eagerly exploring the possibilities.

Engineers envision creating even more sophisticated xenobot designs, while biologists are exploring how to improve the cell cultures and develop cells with advanced functions, such as programmed biodegradability or targeted drug delivery capabilities.

As the field grows, partnerships are extending to incorporate more specialized areas, such as artificial intelligence for predicting xenobot behavior or materials science to enhance cellular durability.

The future of xenobot research will likely involve larger and more diverse teams, each bringing a unique perspective to the development process.

This synergy between biologists and engineers exemplifies a new way of conducting science—where living systems and technology converge to create tools that were once the realm of science fiction.

Their work is transforming our understanding of what constitutes life and robotics, leading us toward a future where biohybrid technologies like xenobots may become commonplace.

This collaboration page captures the exciting fusion of biology and engineering, explaining how each discipline contributes to the pioneering work of xenobot creation, and highlights the dynamic problem-solving approach that characterizes their ongoing efforts.

Highlighting the Importance of Interdisciplinary Teamwork

The creation of xenobots, biological machines that mark a significant leap in synthetic biology and bioengineering, exemplifies the power of interdisciplinary teamwork. Xenobots are living cells configured to perform specific tasks that traditional robots or purely biological cells could not accomplish independently. This convergence of biology, computer science, engineering, and robotics has led to a new realm of programmable organisms with the potential to revolutionize medicine, environmental management, and more. Understanding and creating xenobots required insights and expertise from diverse scientific domains, highlighting the essential role of interdisciplinary collaboration in achieving breakthroughs that no single field could reach alone.

The Pillars of Xenobot Design: Biologists, Engineers, and Computer Scientists

The creation of xenobots called for collaborative efforts between biologists, engineers, and computer scientists, each of whom contributed unique expertise essential to the project's success. Biologists provided fundamental knowledge about cell behavior, physiology, and embryonic development, enabling researchers to manipulate cells in innovative ways. Engineers, particularly those with expertise in bioengineering, offered skills in design and modeling, essential for creating functional structures and ensuring that these living organisms could perform specified tasks effectively. Computer scientists, on the other hand, were crucial in programming these biological organisms, utilizing algorithms to predict optimal designs and behaviors.

Computer simulations were essential in exploring the vast array of possible configurations for xenobots. Algorithms generated models of cellular clusters, simulating how different arrangements of cells might function. This computational phase enabled scientists to narrow down configurations before testing them in the lab, significantly reducing trial-and-error and accelerating the process.

Each phase relied on the integration of biological insights, engineering principles, and computational power, illustrating how different fields collectively paved the path for the development of xenobots.

Shared Vision and Innovation through Diverse Perspectives

One of the most challenging aspects of interdisciplinary research is integrating the unique "languages" and methodologies of each field involved. Biologists, engineers, and computer scientists often approach problems with distinct perspectives. For example, while biologists may focus on understanding cellular interactions and maintaining viable cell cultures, engineers might emphasize structural integrity and mechanical properties. Meanwhile, computer scientists tend to think in terms of algorithms and data models, focusing on computational efficiency and prediction accuracy.

Despite these differences, the xenobot project showcases how a shared vision can align different perspectives to achieve a common goal. Scientists were united by the ambition to create a new form of bio-robot that could complete tasks in a way no existing technology could, like cleaning up microplastics in oceans or delivering targeted drug treatments.

The project benefitted from each field's unique insights, and working in tandem helped the team move beyond individual limitations to solve complex problems holistically. With each phase of the project, interdisciplinary teams were able to leverage the strengths of each discipline, integrating biological knowledge with engineering and computational precision to create entirely new possibilities in the form of xenobots.

Problem-Solving through Interdisciplinary Synthesis

Creating functional xenobots required overcoming challenges that no single field could address on its own. For instance, designing xenobots capable of movement and task completion presented numerous biological and engineering challenges. Biological cells, particularly frog embryonic cells used in xenobot experiments, do not naturally assemble into forms that can perform mechanical tasks. Engineers needed to ensure the structural stability and viability of cell clusters while enabling complex behaviors, like movement toward or away from stimuli. Here, interdisciplinary teamwork allowed the fusion of different expertise to navigate these challenges.

For instance, the project involved utilizing frog stem cells, which required understanding how these cells could be coaxed to self-assemble and behave in a desired manner.

Engineers worked closely with biologists to test configurations for stability, learning how to assemble cells without compromising their integrity.

This effort required continuous communication and problem-solving, as the disciplines adapted their strategies based on one another's findings.

The result was a collaborative, iterative process in which each team brought unique solutions and received feedback from others, enhancing the development of viable, motile xenobots.

Adaptive Learning and Continuous Feedback
Another key advantage of interdisciplinary teamwork in the xenobot project was the role of continuous feedback loops.

As engineers and biologists tested configurations predicted by computer models, they gained new insights that were then shared back with the computational team.

This feedback often led to improved algorithms and simulation models, making future predictions more accurate. Such an iterative cycle of learning and refinement helped optimize the design of xenobots.

This iterative process also highlighted how teams can adapt to evolving challenges through a flexible, interdisciplinary approach.

For example, if a configuration failed to behave as expected in lab tests, it wasn't simply discarded; instead, biologists and engineers analyzed the reasons for its failure, then worked with computer scientists to adjust the simulation parameters.

This adaptive process allowed the entire team to learn from each experiment, reinforcing the benefits of interdisciplinary collaboration and continuously improving both the design and modeling techniques for xenobot development.

Future Directions: Expanding Interdisciplinary Networks

The success of the xenobot project opens doors to expanding interdisciplinary networks further, as researchers consider more complex tasks and applications for bio-robots. In the future, the field may bring in expertise from additional disciplines such as materials science, regenerative medicine, and even environmental science, to explore more specific applications like tissue regeneration, targeted cellular therapies, and ecological preservation. By involving an even broader range of disciplines, future projects may advance xenobot design to new heights, creating tailored solutions for pressing global challenges.

This teamwork approach also highlights a transformative shift in scientific research, where disciplines that once operated in isolation are now working in concert to address complex, multifaceted problems. This trend promises not only to accelerate the pace of innovation but also to open up entirely new fields of study. For instance, the field of biocomputing, which blends biological organisms with computational functions, may yield future applications unimaginable without the collaborative foundation set by xenobot research.

A New Paradigm of Collaboration in Science

The development of xenobots represents more than a technical achievement; it's a powerful demonstration of how interdisciplinary teamwork can drive scientific and technological progress. By uniting experts from diverse fields, the project showcases a new paradigm where collaboration across disciplines is not only beneficial but necessary for pushing the boundaries of what is possible. Xenobots are thus a product of collaborative ingenuity, embodying the collective power of biologists, engineers, and computer scientists working together to transform science fiction into reality.

This approach to teamwork, in which individual expertise is synthesized into collective innovation, will undoubtedly continue to play a vital role in future scientific breakthroughs. As researchers from various disciplines join forces, the potential to solve complex global challenges expands, showing that the best solutions emerge not from a single field, but from a harmonious blend of knowledge, creativity, and vision across disciplines.

How Biologists and Engineers Collaborate to Create Xenobots

The creation of xenobots—the world's first "living machines"—marks a revolutionary intersection between biology and engineering. In these extraordinary collaborations, biologists and engineers combine their expertise to turn living cells into programmable life forms capable of performing specific tasks. This process leverages the natural properties of cells and the precision of computational design, breaking new ground in both science and technology.

Here's a look at how these disciplines intersect to create xenobots and what this synergy entails.

Understanding Xenobots

Xenobots are made from living cells, often sourced from frog embryos, specifically the species Xenopus laevis, which is where their name originates. These cells are harvested and reassembled into new forms using computer algorithms to predict and shape their behaviors.

Unlike traditional robots, xenobots are not made of metal, plastic, or electronic components; they're entirely organic and biodegradable, raising the potential for applications that traditional robots might struggle to perform—such as operating in delicate ecosystems or inside the human body without causing harm.

Defining Roles in the Collaboration

The process of creating xenobots is a truly interdisciplinary endeavor, relying heavily on the specific skills of both biologists and engineers. Biologists contribute their deep knowledge of cell behavior, tissue organization, and the fundamental properties of life.

Engineers, particularly those with expertise in artificial intelligence (AI) and robotics, bring their skills in computational modeling, design optimization, and automation.

These complementary skill sets enable them to solve unique challenges that arise when designing life forms with novel functions.

Biologists' Role

Biologists are responsible for handling and preparing the cells that will eventually form the xenobot. They select the appropriate cell types, often choosing from skin or heart cells from frog embryos, because these cells have unique properties that can be harnessed to create movement or structural integrity. Skin cells offer robustness and cohesion, while heart cells can contract rhythmically, providing a natural form of locomotion. Biologists carefully culture and nurture these cells, ensuring their viability and arranging them in specific patterns according to the computer-generated designs provided by the engineers.

Engineers' Role

Engineers, on the other hand, are responsible for designing the shape, structure, and function of the xenobot before it is physically assembled. They use advanced computer algorithms, typically based on artificial intelligence, to simulate different cell configurations and predict their behaviors. Engineers input parameters related to the task the xenobot is expected to perform—whether it's movement, object transport, or targeted cell manipulation—and the algorithms then run thousands of virtual simulations to determine which structure will be most effective. Once the design is selected, engineers work closely with biologists to communicate the exact cellular arrangements required.

The Computational Side: How AI Drives Xenobot Design

AI plays a pivotal role in creating xenobots by running simulations to generate viable designs. Engineers use algorithms to explore a vast range of potential cell configurations, optimizing each design based on specific performance criteria. The AI considers factors like size, shape, and the arrangement of different cell types to predict which xenobot structure will achieve the desired behavior.

Engineers rely on these simulations because they enable rapid iteration—thousands of configurations can be tested in a matter of hours. Without this computational approach, each design would need to be tested manually, which would be time-consuming and labor-intensive. The computational model streamlines the process, allowing engineers to focus on the most promising designs.

Building Xenobots in the Lab: Where Biology Meets Engineering

Once a design is selected, biologists and engineers move to the laboratory to bring it to life. Using microsurgical techniques, biologists carefully arrange the cells to match the structure suggested by the computational model.

This step requires extreme precision, as each cell must be positioned correctly to allow the xenobot to function as intended.

Biologists use fine instruments, such as micro-pipettes and forceps, to position the cells, effectively "sculpting" them into the desired shape.

Engineers provide support by translating the digital blueprint into practical instructions, sometimes even developing custom tools to assist in the cell arrangement process. It's a meticulous process, but it's essential for achieving a working xenobot.

Testing and Refining Xenobots

Once assembled, xenobots are tested in a controlled environment. They are placed in small dishes or chambers where researchers can observe their behaviors, verifying if they move, perform tasks, or behave as predicted.

This testing phase is crucial because it provides feedback that both biologists and engineers use to refine future designs.

If a xenobot doesn't perform as expected, the team can return to the drawing board.

Engineers may adjust the digital model to alter the shape or cell configuration, while biologists might experiment with different cell types or growth conditions.

This iterative process enables the team to gradually improve the xenobots' capabilities, making each generation more adept at fulfilling its intended purpose.

Challenges in Collaboration: Navigating Different Languages and Approaches

While the collaboration is highly productive, it's not without challenges. Biologists and engineers often speak different scientific "languages" and use different frameworks for problem-solving. Engineers are typically solution-oriented and accustomed to defining clear parameters for their models, while biologists work in the often unpredictable realm of living organisms, where outcomes aren't always controllable.

This can create communication barriers, as each discipline has unique expectations and methods.

To overcome these challenges, teams often develop a shared understanding, holding regular interdisciplinary meetings to align their goals, clarify terminology, and address issues as they arise.

Open communication and mutual respect are key to these collaborations, allowing biologists and engineers to bridge their disciplinary divides and work toward a common objective.

The Future of Xenobot Collaboration

As xenobot research advances, collaboration between biologists and engineers is likely to deepen. Both fields are rapidly evolving, and breakthroughs in cellular reprogramming, gene editing, and bio-robotics could further enhance what xenobots are capable of.

Engineers are exploring more sophisticated AI models to predict complex behaviors, while biologists are discovering new ways to manipulate and program cellular functions.

One exciting possibility lies in creating xenobots that can self-repair, or even self-replicate, which would open doors for applications in medicine, environmental cleanup, and regenerative medicine.

Another frontier involves integrating sensory capabilities, enabling xenobots to react to their environment in real-time, such as changing direction upon detecting specific chemicals or toxins.

A New Age of Living Machines

The creation of xenobots represents a paradigm shift, not just in robotics but in our understanding of life's possibilities.

Biologists and engineers are pioneering a new field of "biological machines" that blurs the line between the natural and artificial, offering tools for tackling some of humanity's most complex challenges.

By merging biology's insights with engineering's precision, these researchers are developing life forms that may transform medicine, environmental science, and beyond, ushering in an era where the collaboration between these disciplines opens up possibilities we are only beginning to imagine.

Chapter 14: Xenobot-Organism Interactions

Xenobots, tiny biological robots built from frog stem cells, hold potential as a revolutionary tool in various scientific and medical applications. They are designed to perform targeted tasks within the body, including removing harmful bacteria or facilitating healing. But how do xenobots interact with natural organisms?

These interactions reveal much about the emerging field where synthetic biology meets traditional biology, opening doors to new therapeutic approaches while also posing unique challenges.

Knowing the Foundation of Xenobot Interactions

To fully appreciate xenobot-organism interactions, we first need to understand the design and functionality of xenobots. Constructed from the embryonic cells of the African clawed frog (Xenopus laevis), xenobots are a unique blend of biological material that behaves in ways that resemble both natural and synthetic processes.

Unlike purely mechanical robots, xenobots use cellular structures and chemical signaling, making their interactions with organisms vastly different from traditional nanotechnology.

Xenobots operate using the inherent programming of their frog cells. These cells naturally possess motility, a form of movement critical to their functions.

The xenobots' behavior is also influenced by cellular communication processes, like signaling pathways that dictate their responses.

By tapping into the natural cellular language, researchers have unlocked the potential for xenobots to operate seamlessly in environments where organic and inorganic factors are at play.

Mechanisms of Interaction

When placed within an environment containing other organisms, xenobots can interact in several ways:

1. Cellular Communication: Just as cells communicate through signaling molecules, xenobots can engage in cellular crosstalk with native cells. This form of biochemical exchange allows them to coordinate actions or influence nearby cell behavior. For example, if xenobots are introduced into a wound site, they can release signaling molecules that promote cell migration and accelerate tissue repair.

2. Physical Interaction: Xenobots can physically interact with cells and small organisms in their environment, gently pushing them or clustering them together. This behavior has significant applications, such as organizing cellular debris or gathering microorganisms.

In experimental settings, xenobots have demonstrated the ability to move toward or away from specific chemical signals, enabling them to approach harmful pathogens selectively.

3. Environmental Sensing: Xenobots can also respond to cues in their surroundings, detecting chemical gradients or temperature shifts. By sensing these environmental factors, they can adjust their behavior, allowing them to locate infections or areas requiring treatment within a biological organism.

This sensing capability is crucial for autonomous behavior and adaptive responses in complex, dynamic environments.

4. Collective Behavior: Interestingly, xenobots can work together, amplifying their interactions with other organisms. When grouped, they exhibit a form of "swarm intelligence," coordinating movements and actions in response to environmental stimuli.

This collective behavior could be useful in large-scale therapeutic tasks, such as clearing blockages in blood vessels or creating cellular scaffolding to support tissue growth.

Therapeutic Applications

One of the most promising applications of xenobot-organism interactions lies in healthcare. Since xenobots are made from biological material, they have the potential to interact with human cells without causing immune rejection, a common issue with synthetic implants or nanoparticles.

They can be designed to perform several tasks within the body, including:

• **Clearing Bacterial Infections:** By moving toward harmful bacteria, xenobots could serve as micro-scale "cleaners," capturing pathogens and transporting them to areas where they can be destroyed. Unlike antibiotics, which may harm beneficial bacteria and lead to resistance, xenobots offer a targeted approach that minimizes side effects.

• **Promoting Healing:** Xenobots can encourage tissue repair by signaling to nearby cells, prompting processes like cell division and migration. Their presence could facilitate healing in wounds, burns, or surgical sites by acting as a support system, ensuring that the body's natural repair mechanisms are optimized.

• **Delivering Targeted Drugs:** Xenobots can carry and release drugs directly at infection or tumor sites, reducing systemic exposure and enhancing treatment effectiveness. This method reduces the risk of side effects associated with many medications and provides a more efficient drug delivery system.

Ethical Considerations and Environmental Impact

While xenobots offer remarkable benefits, their interactions with organisms also bring ethical considerations. For example, if xenobots were to interact with unintended organisms or persist in ecosystems, they might disrupt natural balances. This raises questions about the containment and disposal of xenobots after their tasks are complete. Since they are made from living cells, xenobots will eventually degrade, but understanding their potential ecological impact remains a critical area of research.

In response to these concerns, scientists have developed biocompatible and biodegradable xenobots that naturally break down once their tasks are done. Researchers are also exploring ways to ensure that xenobots target only specific cells or areas, minimizing unintended interactions with non-target organisms.

Case Study: Xenobot Interaction with Bacteria

A compelling example of xenobot-organism interaction involves experiments conducted with bacterial cultures. In laboratory settings, researchers observed that xenobots could be directed to interact with E. coli bacteria. The xenobots detected the presence of bacterial colonies, moved toward them, and eventually encapsulated them, isolating the bacteria from surrounding cells. This interaction is particularly promising for potential applications in targeted infection control within the body.

In trials, xenobots showed an ability to respond not just to E. coli, but to a range of bacterial signals, indicating a capacity for broader anti-bacterial applications.

Moreover, because the xenobots were made of biological tissue, they posed no harm to surrounding non-target cells, a significant advantage over traditional antimicrobial methods.

Challenges in Xenobot-Organism Interactions

Despite their potential, xenobot-organism interactions present technical challenges. Xenobots require precise control, particularly in complex biological environments where numerous cells and microorganisms coexist. Achieving this level of specificity requires a deep understanding of how xenobots respond to diverse biochemical signals, a field still in its early stages.

Additionally, the energy needs of xenobots are an area of ongoing research. While xenobots can draw energy from the environment or from engineered "fuel cells" embedded within their structure, energy management becomes challenging when interacting with organisms over extended periods. Scientists are exploring ways to make xenobots more efficient, such as by enabling them to harvest energy from their surroundings.

The Future of Xenobot-Organism Interactions

Looking ahead, the interaction between xenobots and organisms holds incredible potential.

As research advances, scientists aim to develop xenobots that can seamlessly integrate with natural biological systems, performing highly specialized tasks without adverse effects.

The goal is to create xenobots that are adaptable, responsive, and capable of safely operating within the human body, perhaps even collaborating with human cells to treat diseases at a cellular level.

By understanding and refining xenobot-organism interactions, researchers are paving the way for an era where medical interventions are as small, precise, and personalized as the cellular level itself.

The journey of xenobots is just beginning, but their promise hints at a future where artificial biology and natural biology work hand in hand to improve health, restore balance, and open new frontiers in medicine and environmental care.

Studying the Interaction Between Xenobots and Other Living Organisms

In the evolving landscape of biotechnology, xenobots—tiny, programmable, living robots made from frog stem cells—represent an extraordinary frontier in the study of life and machine.

These biological robots, although small and simple, can be carefully designed to perform a variety of tasks.

Among the most exciting avenues of research involving xenobots is their interaction with other living organisms.

Understanding these interactions not only helps in assessing the biological compatibility of xenobots but also unlocks possibilities for future applications in medicine, environmental conservation, and more.

The Basics of Xenobot-Organism Interactions

Xenobots, while essentially programmable clusters of cells, behave in ways that make them distinctly capable of interacting with their environments. By design, they can respond to various stimuli and carry out tasks like moving toward certain chemicals or substances, which provides a basis for interaction with living organisms.

At the cellular level, xenobots communicate using chemical signals and physical movement. Researchers are interested in how these responses affect and are affected by nearby living cells or organisms, a study area that has sparked interest in various scientific fields, including cellular biology, bioengineering, and ecology.

Communication Through Chemical Signaling

Chemical signaling plays a major role in how xenobots interact with other living entities. All cells communicate using chemical signals, and xenobots are no different. By using chemical gradients or molecular cues, xenobots can "talk" to other cells in their vicinity. When placed in an environment with other living cells or even microorganisms, xenobots can emit or respond to chemical signals in a way that enables mutual interaction.

For instance, xenobots could release certain molecules that either attract or repel other cells, affecting cell migration patterns or colony formations. These interactions are closely monitored in research labs, where scientists analyze how xenobots' signaling might influence bacterial colonies or impact cellular structures within a controlled ecosystem.

This capacity for chemical communication opens a new realm of potential applications, particularly in medicine. Xenobots could one day be used to deliver targeted treatments by recognizing and responding to specific chemical markers in diseased cells, such as tumors.

By interacting at a chemical level, xenobots might even be capable of influencing biological processes like inflammation or cell repair, facilitating more precise medical interventions.

Physical Interactions and Movement-Based Responses

Beyond chemical signaling, xenobots' physical presence and movement also influence how they interact with living organisms. As they move, xenobots can come into direct contact with other cells, creating new kinds of interactions. For example, when xenobots encounter bacterial biofilms, they might disrupt the colony's structure, which could have implications for bacterial control or the prevention of infection.

In some experiments, xenobots have been observed physically herding or gathering small groups of cells, showcasing an intriguing possibility of xenobots as "micro-managers" of cellular environments. This ability to physically influence cellular positioning could be beneficial in wound healing or tissue regeneration, where spatial organization is crucial for recovery. For instance, in a wound environment, xenobots might help align cells to accelerate healing or remove cellular debris, enhancing natural recovery processes.

Immune System Responses and Biocompatibility

A critical aspect of studying xenobot-organism interactions involves understanding how the immune system responds to these biological machines. Although xenobots are made from biological materials, they are still foreign entities to the human body or any other animal they might interact with in real-world applications. The immune response to xenobots is a crucial area of study, as it determines their potential viability for use inside the body. Early experiments suggest that xenobots, due to their cellular makeup, might evade or modulate immune responses, but more research is necessary to confirm this and assess long-term effects.

Biocompatibility is another key focus. Since xenobots are derived from frog cells, they are inherently non-human, raising questions about how compatible they would be with human or other mammalian tissues. Scientists are exploring ways to modify xenobot structures and behaviors to make them more universally compatible across different organisms. For instance, introducing human stem cells into the xenobot structure could theoretically enhance compatibility with human tissue, though this requires careful ethical and regulatory considerations.

Ecological Interactions and Environmental Applications

Outside of the medical sphere, xenobots have potential applications in environmental science, particularly in interacting with microorganisms within ecosystems. For instance, xenobots could be engineered to identify or even neutralize harmful bacteria or toxins in water sources.

These microbots could interact with bacteria or pollutants, either binding to or breaking down harmful substances, thereby serving as biological cleanup agents. The ability of xenobots to target and manipulate specific organisms in an environment could be particularly useful in maintaining ecological balance, especially in polluted or sensitive habitats.

However, these interactions also raise ecological concerns. The introduction of xenobots into natural ecosystems must be carefully managed to prevent unintended consequences.

Researchers are studying how xenobots affect microbial communities and how they might interact with various species across different environments, aiming to ensure that any environmental deployment of xenobots is safe and sustainable.

Social and Ethical Implications

The potential for xenobots to interact with living organisms on multiple levels also raises ethical questions about their impact on life and ecosystems. Given that xenobots are living organisms in their own right, albeit artificially constructed, scientists and ethicists must consider their rights, if any, and the moral implications of deploying them in settings that involve other living beings.

Furthermore, there is the broader question of control and predictability: while xenobots can be programmed for specific tasks, the complexity of living systems means there is always a degree of unpredictability in their interactions.

Studies on xenobot interactions are, therefore, also dedicated to understanding possible risks and ensuring safe and controlled applications.

Future Directions in Xenobot-Organism Interaction Studies

As the field of xenobot research progresses, scientists are increasingly exploring ways to refine xenobot interactions with other organisms for specific, targeted outcomes.

Through advanced programming, xenobots could be made more responsive to specific cues, allowing for sophisticated interactions with various biological systems.

Researchers envision xenobots capable of delivering medicine directly to infected cells, repairing tissues, or even interacting symbiotically with human cells for enhanced therapeutic effects.

In sum, the study of xenobot-organism interactions is a rapidly evolving area, and it holds the potential to transform fields as diverse as medicine, environmental science, and ethics.

As we uncover more about how these biological machines communicate and interact with life around them, we not only expand our understanding of life's building blocks but also lay the groundwork for a future where biology and technology seamlessly integrate for the benefit of all.

Xenobots as Tools for Studying Ecosystems and Behavior

In recent years, xenobots—tiny living robots crafted from biological cells—have emerged as revolutionary tools in the study of ecosystems and behavior.

Developed through a fusion of biology and engineering, xenobots are built using the cells of the African clawed frog (Xenopus laevis), but these living organisms are designed to perform functions that natural cells typically do not.

By leveraging the malleability of biological tissue with the precision of engineered design, scientists can now create xenobots capable of interacting with the environment in ways that can reveal new insights into ecological dynamics, cellular communication, and even species behavior.

Xenobots are not robotic in the traditional sense; they lack metal, circuits, and wires. Instead, they are organic entities, created entirely from living cells and capable of movement, sensing, and interacting with each other and their environment.

Because they are made from frog cells, xenobots have unique advantages over traditional artificial robots when studying ecosystems. These advantages include environmental compatibility, biodegradability, and an ability to adapt to their surroundings, making them ideal candidates for ecosystem research.

Comprehending Ecosystem Dynamics with Xenobots

One of the primary ways xenobots contribute to ecosystem research is by allowing scientists to study microenvironments and nutrient cycles. Ecosystems are complex webs of interdependent species, environmental factors, and energy flows.

Even at a micro-scale, these systems are often difficult to study due to limitations in current robotic technology. Traditional robots are often too large or invasive, disrupting the very systems they aim to study.

Xenobots, however, can be introduced into ecosystems without causing significant disturbance. They can navigate water, soil, or microhabitats, allowing them to explore spaces where larger robots cannot venture without interference.

Since xenobots are biologically derived, they do not pose the same risk of contamination that metal-based or plastic-based robots do. Their organic composition makes them less likely to affect the chemical or biological makeup of the environments they explore. This property is especially beneficial in fragile ecosystems, such as wetlands, coral reefs, or marshes, where external interference can have drastic consequences.

Xenobots could potentially assist scientists in studying algae blooms, microbial communities, or soil health without adding pollutants or physical disturbances, allowing for more natural observations and measurements.

Another promising application lies in monitoring pollution levels and the impact of human activities on small ecosystems. Xenobots can be modified to detect changes in pH, temperature, or specific chemical compositions. For example, xenobots could be engineered to fluoresce or change color in response to particular toxins or pollutants, serving as bio-indicators in real time.

By deploying xenobots in water bodies, scientists could gain insights into pollution spread and its effects on biodiversity, providing early warning signs of environmental degradation.

Mapping and Monitoring Behavior of Species
Xenobots also open new doors for studying the behavior of small organisms, which are often difficult to observe in their natural habitats. Unlike most robotics used in field biology, xenobots are small enough to interact directly with microorganisms, enabling them to mimic or respond to natural cues in their environment. This ability allows scientists to observe phenomena like predation, symbiosis, and communication among organisms on a micro scale.

For example, scientists could use xenobots to study predator-prey interactions at a cellular level. By programming xenobots to mimic the movement or chemical cues of certain organisms, researchers can observe how prey species react to potential threats or how predators respond to prey-like movements. Such experiments can be done in laboratory-controlled microhabitats that simulate natural conditions, allowing researchers to observe complex behaviors without having to intervene in actual ecosystems.

Beyond individual behavior, xenobots have the potential to reveal insights into social or collective behavior. Xenobots can be designed to function as individual units, or they can be engineered to work together, similar to how ants, bees, or other social organisms operate. By programming xenobots to communicate with each other or to respond to particular stimuli, scientists can use them to study collective decision-making, group movement, and resource-sharing behaviors. The insights gained from these experiments could have applications not only in understanding natural behaviors but also in developing collective artificial intelligence systems.

Studying Cellular Communication and Behavior

On a microscopic level, xenobots also offer unprecedented opportunities for studying cellular behavior and communication. Cells communicate through a complex system of chemical signals, and by altering the chemical environment or introducing xenobots with specific signaling molecules, researchers can observe how cells respond and adapt. These studies may shed light on how organisms organize themselves at a cellular level, how they repair damage, or how they defend against pathogens.

Additionally, xenobots can help scientists understand tissue regeneration and healing processes. By programming xenobots to imitate wound-healing mechanisms or cell migration patterns, researchers can explore how cells coordinate to repair damaged tissue. This understanding could have far-reaching implications, including the development of new therapeutic techniques for human medicine, where cellular behavior is critical for health and recovery.

Xenobots and Ethical Implications in Ecosystem Research

While xenobots present exciting opportunities, their use in ecosystems raises ethical and environmental concerns. As living organisms engineered by humans, xenobots represent a novel form of life. Their deployment into natural environments must be carefully regulated to avoid unintended consequences. There is ongoing debate within the scientific community regarding whether xenobots, as living constructs, might impact existing ecosystems in unforeseen ways, potentially competing with native organisms or influencing ecological balance.

To address these concerns, scientists are working to ensure that xenobots are biodegradable and short-lived. The lifespan of a xenobot can be controlled, and they are designed to break down naturally over time. Additionally, researchers are exploring ways to implement self-destructive programming, allowing xenobots to cease functioning after a set period.

These safeguards aim to reduce the ecological impact and ensure that xenobots do not proliferate or disrupt ecosystems once their scientific purpose has been fulfilled.

The Future of Xenobots in Ecosystem and Behavioral Studies

The use of xenobots in ecosystem research is still in its infancy, but the potential applications are vast.

As technology advances, scientists may develop xenobots that are even more sophisticated, capable of performing increasingly complex tasks in diverse environments.

For example, future xenobots could be equipped with sensors for real-time data collection, providing researchers with a wealth of information about environmental conditions and organism behavior in situ.

Ultimately, xenobots offer a glimpse into a future where biology and engineering merge to create tools that can deepen our understanding of ecosystems and behaviors in ways never before possible.

They represent an innovative, flexible, and eco-friendly approach to studying nature at its most fundamental levels, bridging the gap between living systems and human technology.

As scientists continue to refine xenobot technology, they may unlock new pathways for understanding the delicate balance that sustains life on Earth, helping us protect and preserve our ecosystems for future generations.

Chapter 15: Xenobot Ethics and Regulations

The emergence of xenobots—biological robots crafted from living cells—has brought about thrilling possibilities for science, medicine, and environmental conservation. These tiny, programmable organisms offer a future where bioengineered constructs could help repair damaged tissues, clean up environmental waste, and perform a range of tasks that were once the realm of science fiction.

However, these groundbreaking developments raise significant ethical and regulatory questions. How can we ensure that xenobots are used responsibly? What protections need to be in place for both the environment and society?

In this chapter, we will explore the ethical implications of xenobots and the frameworks needed to govern their development and application.

Ethical Dimensions of Xenobot Technology

As xenobots become more advanced, they challenge our traditional notions of biology and robotics. Unlike conventional robots, xenobots are made from living cells. This distinction brings them closer to life forms than machines, making ethical considerations complex and unique.

Here are some of the primary ethical concerns surrounding xenobot technology:

1. Defining Life and Responsibility

One of the fundamental ethical questions surrounding xenobots is whether they should be classified as living beings or as tools. While xenobots lack the hallmarks of life as we understand it—such as the ability to reproduce independently, feel pain, or possess consciousness—they are still derived from living cells and have a limited capacity for self-repair.

The question arises: Should we treat them as entities deserving protection or merely as disposable tools? This classification impacts not only how we regulate xenobot production but also how we approach their use and disposal.

2. Risk of Unintended Consequences

Xenobots are created with specific functions in mind, such as navigating a particular environment or delivering targeted therapeutic agents. However, their interactions with other organisms or their responses to environmental changes are not always fully predictable.

Ethical concerns arise regarding unintended consequences—such as the possibility that xenobots could mutate, cause unforeseen ecological effects, or interact with existing ecosystems in harmful ways. How do we responsibly manage the deployment of xenobots to minimize risks of unintended consequences?

3. The Role of Artificial Intelligence (AI) in Xenobot Development

The design and programming of xenobots often rely on AI-driven algorithms that can produce novel, optimized shapes and functions. AI brings its own set of ethical concerns, particularly around transparency and accountability. If an AI algorithm designs a xenobot that behaves unexpectedly or causes harm, who bears the responsibility? The integration of AI also raises questions about control and predictability, as it's difficult to fully understand the decision-making processes within complex algorithms.

4. Human Enhancement and Control over Nature

Xenobots symbolize an unprecedented level of control over biological matter, which brings up ethical questions about human enhancement and interference with nature. With the ability to manipulate life forms for specific tasks, are we crossing a line into the realm of "playing God"?

Many ethicists argue that while such technologies hold promise, they also require a respect for natural processes and a commitment to avoid hubris. Balancing innovation with a respect for natural ecosystems is key to developing xenobots in a responsible way.

5. Biocompatibility and Safety Concerns

If xenobots are to be used in human healthcare—for example, to deliver drugs within the body or perform microsurgery—ensuring their biocompatibility and safety is essential. Ethical guidelines are necessary to determine how xenobots can be introduced into the human body and what safeguards are needed to protect patients from adverse effects.

Protocols for safe disposal of xenobots, both in medical and environmental settings, must also be developed to prevent contamination or unintended interactions with other organisms.

Regulatory Frameworks and Governance

In light of these ethical considerations, a regulatory framework is necessary to guide the safe and responsible development of xenobot technology.

Effective governance would need to address the following areas:

1. Research Oversight and Approval

Xenobot research should be subject to rigorous oversight to ensure ethical practices and adherence to established standards. Institutional review boards (IRBs) and ethics committees should be involved in assessing xenobot research proposals, especially when human or animal subjects are involved. In cases where xenobots are introduced into natural environments, approval processes should require thorough ecological risk assessments.

2. Guidelines for Transparency and Accountability
Transparency in xenobot development is crucial to public trust and responsible science. Regulatory bodies should establish guidelines for documenting and sharing information about xenobot design, testing, and deployment processes. Open disclosure of potential risks, research methods, and funding sources can help build accountability. Additionally, regulatory frameworks should clarify who is responsible if xenobots cause harm—be it the developers, manufacturers, or institutions overseeing their deployment.

3. Biosafety Standards and Environmental Impact Assessments

Biosafety protocols are essential to managing xenobot technology, especially when xenobots are introduced to external environments.

Just as genetically modified organisms (GMOs) are subject to strict containment and testing standards, xenobots should be governed by similar, if not more stringent, biosafety guidelines.

Regulatory frameworks could mandate that any xenobot deployment include an environmental impact assessment, which evaluates the potential consequences of introducing xenobots to various ecosystems.

4. Establishing Ethical Limits on Xenobot Capabilities

Regulatory frameworks should establish limits on xenobot capabilities to prevent misuse or potentially harmful applications.

For instance, strict guidelines could prohibit the development of xenobots capable of uncontrolled self-replication or possessing complex AI autonomy.

Clear boundaries can help prevent ethically questionable uses of xenobots while allowing for productive scientific advancements.

5. Protocols for Xenobot Disposal and Lifecycle Management

Xenobots require special protocols for safe disposal once they have completed their tasks.

Ensuring that xenobots degrade safely, without leaving lasting impacts on the environment, is crucial.

Guidelines for lifecycle management, including biodegradability standards, can help protect ecosystems from contamination.

Proper disposal practices can also address concerns about xenobots persisting in unintended environments and potentially causing harm.

6. Global Collaboration and Standards

Given the potential international impact of xenobot technology, global collaboration is essential. Countries should work together to establish universal standards and guidelines that govern the ethical use of xenobots. International bodies, similar to those overseeing nuclear and biohazard protocols, could be developed to manage xenobot research and application globally.

This approach would help prevent regulatory "loopholes" where xenobot research in one country could affect ecosystems or populations in another.

Public Engagement and Education

Finally, a vital component of responsible xenobot development involves engaging the public. As xenobot technology progresses, it's essential to keep the public informed about potential benefits, risks, and regulatory safeguards.

Encouraging open dialogues and involving communities in decision-making can increase public understanding and acceptance, as well as help refine ethical frameworks.

Transparent communication about xenobot research, especially concerning safety and environmental impact, can foster an informed and engaged society that participates actively in the governance of new technologies.

Xenobots hold incredible promise, but their potential also raises questions that extend beyond the laboratory.

Addressing these ethical and regulatory concerns proactively allows us to shape a future where xenobots can be used safely and responsibly.

As the field evolves, so too will the ethical and regulatory landscape, adapting to new discoveries and challenges in the ever-evolving intersection of biology and technology.

Overview of Regulatory Frameworks for Xenobot Research

As xenobots—bioengineered, self-replicating organisms made from living cells—advance, regulatory frameworks become crucial to ensure these innovations are safely integrated into society.

Regulatory guidelines establish boundaries, mitigate risks, and ensure ethical and environmental protections.

Because xenobots combine biological tissue with computational design, they challenge current regulatory structures, which have traditionally addressed either living organisms or artificial devices but rarely both.

1. Defining Xenobots and the Need for New Regulatory Frameworks

Xenobots differ from traditional synthetic biology due to their self-replicating and programmable capacities. Regulatory bodies thus face unique challenges in categorizing and overseeing these entities.

With features of both biological organisms and robotic devices, xenobots fall at the intersection of biotechnology, artificial intelligence, and robotics regulation.

Traditional frameworks in biotechnology focus on genetically modified organisms (GMOs), which differ fundamentally from xenobots, as xenobots are not genetically modified but bio-fabricated using natural cells programmed through computational models.

Since xenobots are designed to perform specific tasks like environmental remediation or targeted drug delivery, regulators must consider factors such as biosafety, ecological impact, and potential ethical issues in controlling their functions.

The ability of xenobots to replicate, adapt, or evolve also calls for new guidelines on containment, tracking, and deactivation mechanisms in the event of unintended consequences.

2. Current Regulatory Structures and Their Limitations

Several regulatory bodies around the world oversee aspects of biotechnology, each with its own standards, but these regulations may not fully encompass xenobot technology. In the United States, the Food and Drug Administration (FDA), Environmental Protection Agency (EPA), and U.S. Department of Agriculture (USDA) have jurisdiction over various biotechnology products. However, xenobots challenge these boundaries due to their hybrid nature.

For example:

- **FDA:** Oversees biomedical products, including biologics, drugs, and devices. Xenobots intended for medical applications (such as drug delivery or surgical assistance) might fall under FDA's purview. However, questions remain on whether they should be regulated as drugs, devices, or biological products, due to their complex living structure and computational components.

- **EPA:** Typically regulates genetically modified organisms released into the environment, focusing on ecological safety and environmental health. Xenobots used in environmental contexts, such as pollutant cleanup, could fall under EPA jurisdiction. The agency's focus on GMOs, however, complicates the inclusion of xenobots, as they are not genetically modified but instead bio-assembled.

- **USDA:** Oversees agricultural biotechnology, particularly genetically engineered plants and animals. Although xenobots are not directly involved in agriculture, USDA regulations might intersect if xenobots are deployed for soil health or pest control.

In the European Union, the European Medicines Agency (EMA) and European Food Safety Authority (EFSA) regulate biotechnology, with strict protocols on GMOs and clinical products. However, these entities also lack clear guidelines specifically tailored for bioengineered organisms like xenobots, and there is ongoing debate about where xenobots fit within existing GMO regulations. Given these limitations, most countries lack a single, coherent regulatory framework for xenobots, leading to fragmented oversight and regulatory gaps.

3. Ethical and Biosafety Considerations

Regulatory frameworks must address ethical concerns and biosafety risks inherent in xenobot research.

Some key considerations include:

• **Containment and Control:** A significant aspect of xenobot regulation is ensuring that these organisms do not inadvertently escape into uncontrolled environments. Biosafety protocols could mandate xenobot "kill switches" or environmental degradation mechanisms that deactivate the xenobots outside of designated areas.

• **Potential Ecological Impacts:** Because xenobots interact with biological systems and may influence ecosystems if released into the environment, regulators must assess their potential ecological impact carefully. Xenobots introduced for environmental cleanup, for instance, should be evaluated for unintended consequences on local flora and fauna.

• **Ethical Frameworks:** Ethical guidelines aim to address public concerns about the creation of artificial life and the manipulation of natural cells for utilitarian purposes. Internationally, ethical concerns over xenobot research align with broader discussions on synthetic biology, highlighting the need for transparency, public input, and responsible innovation. Public acceptance of xenobots might hinge on ensuring that they are used to address critical global challenges rather than trivial or profit-driven purposes.

4. International Cooperation and Emerging Guidelines

International cooperation is essential to create consistent regulatory standards for xenobot research. The Organisation for Economic Co-operation and Development (OECD) and World Health Organization (WHO) have issued general recommendations on synthetic biology and biotechnology, which may serve as templates for xenobot guidelines. Additionally, recent efforts by the Convention on Biological Diversity (CBD) have considered the environmental implications of synthetic biology, potentially covering xenobot applications.

A notable recent development is the Geneva Convention on Artificial Organisms, a proposed international treaty aiming to create universal standards for bio-fabricated organisms. This treaty would cover safety, ethical, and ecological concerns, facilitating information-sharing and standardization across nations. While it has yet to be ratified, such frameworks signal a growing recognition of the need for global governance in this area.

5. Prospects for Adaptive Regulation

As xenobot technology advances rapidly, adaptive regulatory frameworks that evolve with the field may be necessary. Traditional regulatory models are often too rigid to keep pace with technological developments. Adaptive regulation, by contrast, involves iterative policy adjustments based on feedback from ongoing research, public response, and environmental monitoring.

One model for adaptive regulation is the "sandbox" approach, which has been used in fields like financial technology. A regulatory sandbox allows researchers to experiment under close regulatory supervision, testing new technologies within defined limits.

For xenobot research, a sandbox could enable regulators to monitor xenobot behavior, assess risks, and modify guidelines based on observed outcomes, providing a flexible way to address uncertainties in a controlled setting.

6. Future of Regulatory Frameworks for Xenobots

Looking forward, regulatory frameworks for xenobot research will likely continue evolving as our understanding deepens and public concerns shift. Increased collaboration between scientists, policymakers, ethicists, and the public is crucial to ensure that regulations balance innovation with safety and ethical responsibility.

Effective oversight of xenobots will not only protect human and environmental health but also foster public trust in these new technologies, enabling their responsible integration into society.

By combining transparency, adaptability, and robust safety mechanisms, regulatory frameworks for xenobot research can lay the foundation for a future where xenobots contribute to solutions for environmental, medical, and industrial challenges.

However, the path forward will require ongoing dialogue, vigilance, and a commitment to ethical stewardship. Through careful regulation, we can harness the potential of xenobots while respecting the biological and ethical boundaries that guide scientific progress.

Establishing guidelines to ensure responsible xenobot development.

As science fiction steadily transforms into science reality, humankind finds itself at the cusp of an era where living, programmable robots are no longer just the stuff of imagination.

These unique bio-robots, known as "xenobots," are formed from living cells (such as skin and heart cells from frogs) that scientists have carefully reconfigured to act in programmable ways.

Unlike conventional robots made from metals or plastics, xenobots are fully biological, biodegradable, and possess the ability to self-heal, making them an appealing option for a range of future applications, from medical interventions to environmental clean-up. But with such groundbreaking technology comes the critical task of guiding it responsibly.

1. Ethics and Intent: Defining Purpose and Scope

Responsible development begins with understanding and defining why xenobots are created in the first place. To avoid reckless experimentation, xenobot projects should align with clearly outlined purposes, such as improving human health, aiding environmental restoration, or advancing basic scientific knowledge.

Regulatory bodies and ethical review boards must oversee these projects to ensure they align with societal values and prevent projects that might promote harm or misuse.

Just as with gene editing or artificial intelligence, researchers must determine if potential benefits outweigh ethical concerns. Xenobots designed to deliver medicine to targeted areas in the human body, for instance, could improve outcomes in patients with hard-to-reach tumors. Yet, designing xenobots solely for invasive or covert surveillance would likely face ethical objections. Setting clear boundaries on permissible uses of xenobots will help prevent misuse and promote positive applications.

2. Transparency in Research and Development

Transparency is essential for building public trust in xenobot technology. Research findings, methodologies, and potential outcomes should be accessible to the scientific community and the public. Openness about both the capabilities and limitations of xenobots can help mitigate unrealistic expectations and reduce public anxiety about their use.

Transparency also extends to the communication of risks. Scientists and developers must openly discuss possible unintended consequences, such as ecological disturbances or health risks. By maintaining an honest dialogue with the public and policy-makers, developers can ensure that xenobot projects receive informed support rather than suspicion or backlash.

3. Risk Assessment and Containment Protocols

Xenobots, as biological entities, present unique risks compared to traditional robots. Since they are alive, xenobots can potentially replicate, evolve, or interact unpredictably with natural ecosystems. Therefore, any xenobot release into the environment or use within living organisms requires rigorous risk assessment and containment protocols.

Developers must assess the potential ecological impacts of xenobots before deployment. For instance, would xenobots designed to clean up ocean plastics inadvertently harm marine life? Scientists should consider a range of scenarios, including accidental release, unintended replication, or interactions with other organisms, and plan accordingly. Containment protocols might involve designing xenobots with genetic "kill switches" that trigger self-destruction under certain conditions, reducing the risk of unwanted persistence in the environment.

4. Biodegradability and Environmental Impact

Since xenobots are biological, they inherently offer a more sustainable alternative to traditional robots that rely on metals and plastics. Nevertheless, a critical guideline for responsible xenobot development is ensuring that they remain biodegradable and eco-friendly.

Xenobots should be designed with a focus on minimal environmental footprint, breaking down naturally without leaving toxic residues.

To enhance biodegradability, scientists should prioritize materials and cellular components that decompose quickly.

Additionally, ensuring xenobots have a short lifespan could prevent long-term environmental buildup, which can be particularly important for xenobots used in outdoor or large-scale applications like pollution control.

Continuous studies on environmental impact are crucial, and each generation of xenobots must be evaluated to assess their compatibility with natural ecosystems.

5. Incorporating Safety Mechanisms: Genetic and Physical Safeguards

Built-in safety mechanisms are fundamental to prevent xenobots from behaving unpredictably. While genetic "kill switches" can be employed to deactivate xenobots in emergencies, researchers are also exploring physical designs that limit xenobot autonomy, ensuring they fulfill their intended purpose without deviation.

A genetic kill switch might activate when a xenobot encounters certain environmental triggers, such as changes in temperature or chemical composition, which would effectively dissolve or deactivate the xenobot.

Likewise, developers could use "control genes" that limit xenobot life spans or growth capacities. This makes certain that xenobots cannot self-replicate beyond control or evolve independently, thus maintaining safety and predictability.

6. Addressing Privacy and Security Concerns

The potential applications of xenobots in fields like medicine or information gathering raise privacy and security issues. Xenobots could hypothetically be engineered to collect data within the human body or even in sensitive environments. To prevent misuse, developers must adopt stringent data security standards and privacy guidelines.

Privacy protocols should regulate xenobots designed to operate in personal or private spaces, ensuring they do not inadvertently collect or transmit sensitive information. Additionally, cyber security protections should be implemented to prevent hacking or tampering, particularly for xenobots used in medical applications where personal health data might be at risk.

7. Regulation and Oversight: International Standards and Collaboration

Given the global nature of scientific research, xenobot development should follow universally recognized standards. Establishing international guidelines could help prevent unethical practices or unregulated xenobot proliferation. A centralized governing body could regulate xenobot use, ensuring it aligns with the values of safety, sustainability, and societal benefit.

Collaboration between countries, universities, and private sectors will foster uniformity in standards, making it easier to monitor and control xenobot applications. International cooperation can also facilitate swift responses to emergencies or unintended consequences, ensuring that xenobot technology is shared responsibly.

8. Social and Cultural Sensitivity: Public Engagement and Inclusion

Since xenobot technology affects all sectors of society, its development must include diverse perspectives. Engaging the public through educational campaigns and open forums can help raise awareness and allow citizens to voice their concerns or interests.

Scientists and engineers should actively seek input from communities who may be affected by xenobot applications, especially in cases where they are used for environmental purposes or health interventions.

Creating opportunities for public involvement will foster a culture of shared responsibility and trust, ensuring that xenobot technology aligns with the values and needs of society.

A Path Forward for Responsible Xenobot Innovation

The development of xenobots is a profound step in synthetic biology, holding immense potential for advancements in health, ecology, and beyond. However, as with any powerful technology, the responsibility to regulate and guide its growth lies with scientists, regulators, and society as a whole.

By prioritizing ethical guidelines, transparency, environmental protection, and public trust, we can ensure that xenobots contribute positively to our world.

Following these guidelines will not only safeguard humanity's relationship with this technology but also unlock the potential of xenobots to address some of our most pressing challenges in sustainable, responsible, and innovative ways.

Chapter 16: Xenobots in Education and Research

In recent years, the development of xenobots has captivated scientists, educators, and students alike, creating new opportunities for learning and discovery. Xenobots are the world's first living robots, made from frog (Xenopus laevis) cells and assembled in unique shapes to perform simple functions. These cellular "robots" can move, work together, and even repair themselves to some extent.

Their introduction marks a revolutionary intersection of biology, engineering, and artificial intelligence, promising rich applications not just in science but also in education and research.

The Educational Potential of Xenobots

Xenobots offer a compelling hands-on way for students to engage with concepts in biology, robotics, and bioengineering. By observing these living cells in action, students can witness biology beyond static diagrams and textbook explanations.

Instead of merely reading about cell structures or the concept of self-assembly, students can actively explore these ideas by studying how xenobots are created, how they move, and how they perform tasks like carrying tiny particles.

Fostering Curiosity and Critical Thinking

The curiosity sparked by xenobots can drive students to ask important questions about life, technology, and ethics. What differentiates living organisms from machines? How can living cells be programmed to behave in certain ways? Such questions invite students to think critically, exploring ideas that push traditional definitions of life and artificial intelligence. This engagement with profound scientific questions encourages students to analyze and form reasoned conclusions, a critical skill in scientific research and beyond.

Bridging Interdisciplinary Learning

Xenobots provide a unique teaching tool for interdisciplinary learning. Since creating xenobots involves aspects of biology, chemistry, computer science, and engineering, it presents an opportunity to integrate multiple subjects in the classroom. Teachers can use xenobots as a framework to introduce programming and machine learning in a biological context, or to explore cellular biology alongside principles of robotics.

This interdisciplinary approach can foster a deeper understanding of how various fields of science and technology interact and support each other.

Hands-On Experience and Experimentation

For high school and college students, xenobots can be a fascinating way to gain hands-on experience in the lab. Students can observe how cellular behaviors change based on different structures and environments, experiencing the excitement of working with live materials in real-time. Many students learn best through experiential activities, and xenobots provide an opportunity for interactive learning. They also allow teachers to introduce important research concepts, such as hypothesis testing, experimentation, and data analysis.

Xenobots in Academic Research

Beyond their role in education, xenobots are opening new pathways in scientific research. They offer a platform to study complex biological processes and test hypotheses in ways that were previously impossible.

Investigating Cell Behavior and Communication

One of the most fascinating aspects of xenobots is that they help researchers study cell behaviors in a controlled yet dynamic setting. Xenobots are essentially assemblies of cells that perform tasks, meaning they must communicate and coordinate like cells in living tissues.

By observing how xenobots accomplish coordinated movement or divide tasks among themselves, researchers can gain insights into cellular communication, organization, and cooperation.

This study of cellular interactions is particularly valuable for fields like regenerative medicine, where understanding how cells work together to heal tissues could have a significant impact on health outcomes. Xenobots could also help researchers understand how cells signal each other to form tissues, giving insights into developmental biology.

Testing Genetic and Chemical Influences on Cells

Xenobots offer a new model for testing how cells respond to different chemical and genetic environments. Researchers can modify xenobot cells in various ways to observe how they adapt to changes in their environment. For example, scientists can experiment with different proteins or chemicals to see how these affect xenobot behavior.

They might also explore how genetic modifications influence a xenobot's shape or function.

These studies can offer insights relevant to understanding diseases, particularly those that involve cellular malfunctions, such as cancer. Since xenobots are made from real cells, they respond to genetic and chemical signals in ways that can closely mimic how actual living cells might react in different contexts.

Researchers can use this data to study cellular behavior without the ethical complications of experimenting on living organisms.

Bioengineering and Synthetic Biology

Xenobots represent a living model of synthetic biology, where biological systems are engineered to perform specific functions. Research into xenobot design and optimization is advancing our understanding of bioengineering. This research allows scientists to explore potential applications for self-assembling materials and programmable biological devices.

The more we learn about how to design functional, living robots, the closer we move to breakthroughs in regenerative medicine, bio-computing, and even environmental solutions, such as creating xenobots to remove microplastics from oceans.

Machine Learning and Robotics Advancements

Xenobots are unique in that their structure and movement are largely influenced by machine learning algorithms, which model different configurations to determine optimal designs.

Researchers use computer simulations to test different xenobot shapes and movements before actualizing them in the lab.

This approach provides feedback for developing better artificial intelligence and machine learning systems, particularly in robotics.

Xenobots thus represent a symbiosis between biological systems and artificial intelligence. For instance, researchers can analyze the trial and error of xenobot movements in simulations to improve the algorithms, making this a mutually beneficial field of study.

By studying how artificial intelligence can improve biological systems, scientists and engineers can enhance the precision and adaptability of robotics technology.

Ethics in Xenobot Research and Education

As educational and research tools, xenobots also invite discussions around the ethics of bioengineering. Should we build lifeforms with specific functions, even if they are made from simple cell structures? What distinguishes a xenobot from other organisms, and what responsibilities do we have as creators of new life forms?

These questions are not only important for the scientific community but are also valuable for students who are exploring biology and technology.

Incorporating ethical discussions into the study of xenobots encourages students and researchers alike to consider the societal and environmental implications of their work.

As students learn about xenobots, they also learn to consider the broader impact of scientific innovations on society and nature.

Xenobots are transforming education and research, bridging the gap between the natural and artificial worlds. They provide a gateway for students and researchers to explore a wide range of scientific and ethical questions while offering practical, hands-on experience in an emerging field. Through xenobots, we can inspire curiosity, deepen scientific understanding, and push the boundaries of what is possible with bioengineering.

Whether it's through the classroom or the lab, the study of xenobots is setting the stage for a future where the synergy between biology and technology could lead to breakthroughs in medicine, environmental science, and beyond. As we continue to explore the potential of these living robots, xenobots promise to remain an influential and thought-provoking field for years to come.

Xenobots as Educational Tools for Teaching Biology and Robotics

Imagine a world where education in biology and robotics isn't confined to the classroom or traditional learning materials, but comes alive through tiny, programmable living machines. These "living robots," called xenobots, are not science fiction. They're pioneering creations at the intersection of biology and artificial intelligence, and they offer a groundbreaking tool for education. Xenobots—synthesized from the cells of frogs—are capable of performing programmed tasks like movement, object manipulation, and even self-repair. They present a powerful opportunity to bridge biology and robotics education in ways that are engaging, interactive, and reflective of real-world applications.

Xenobots are named after the African clawed frog, Xenopus laevis, from which they are derived. Developed by researchers using frog stem cells, xenobots are a new class of biological robots, about a millimeter wide, that are neither fully machine nor fully organism. These cell-based, programmable "robots" can be directed to perform specific tasks, such as moving through fluid environments, pushing small particles, or even working together in swarms.

Because xenobots are living cells, they can heal themselves if damaged and degrade naturally when no longer needed, which makes them an environmentally friendly innovation as well.

Their hybrid nature—half biology, half robotics—makes xenobots an incredible educational tool. In classrooms, they can serve as a tangible example of biological engineering, where students can see the blending of natural and artificial systems up close.

In this way, xenobots allow students to explore the basics of biology and robotics through hands-on activities that are both novel and scientifically enriching.

Teaching Biological Concepts with Xenobots

Xenobots provide a unique lens through which students can study fundamental biological processes. Traditional biology education often relies on static models or simulations to teach concepts like cell structure, tissue organization, and the behaviors of living organisms. Xenobots, however, offer a dynamic, interactive way to explore these principles in real time.

Here's how:

1. Cell Biology and Differentiation: Xenobots are built from frog stem cells, which are a type of undifferentiated cell capable of developing into various cell types. By working with xenobots, students gain insight into how stem cells can be programmed to grow into specific forms and perform certain tasks.

They can observe firsthand how clusters of cells can organize themselves, differentiate, and respond to environmental cues, reinforcing the concept of cellular development.

2. Tissue Engineering: Another important aspect of biology that xenobots can teach is tissue engineering. As students observe or interact with xenobots, they can learn about how cells can be arranged in different ways to form functional structures. This is the basis of tissue engineering, a field that has applications in regenerative medicine.

Xenobots demonstrate that cells can be reconfigured to act in new and surprising ways, leading to discussions about potential applications in tissue repair and organ regeneration.

3. Self-Healing and Adaptability: One of the most exciting properties of xenobots is their ability to self-repair. If a xenobot is cut or damaged, the cells can reorganize and mend themselves.

This introduces students to the concept of self-healing in biological systems, which can inspire discussions about similar capabilities in other organisms and potential applications in healthcare and engineering.

4. Bioethics: The use of xenobots in the classroom can also serve as a platform for bioethical discussions. Since xenobots are made from living cells and capable of rudimentary behaviors, questions arise about the ethical considerations of creating and controlling such entities.

Engaging students in these discussions helps them understand the ethical responsibilities that come with biotechnological advancements.

Integrating Robotics and Programming with Xenobots

While xenobots may seem more biological than robotic, their programmability is a perfect introduction to core robotics concepts.

By learning how to instruct xenobots to perform different tasks, students gain insights into programming, control systems, and robotics engineering.

Here are a few examples of how xenobots contribute to robotics education:

1. Algorithm Design: Xenobots can be programmed to follow certain paths, interact with objects, or even work collectively to accomplish tasks. Designing these behaviors requires understanding how algorithms are created and executed.

Students can experiment with different programming logic, modifying the xenobots' behavior by changing input commands and observing the results.

This hands-on approach simplifies the often-intimidating task of understanding how algorithms function.

2. Behavioral Robotics: Unlike traditional robots, xenobots don't have electronic parts or silicon-based processors. Instead, they respond to physical cues, like changes in their environment or the arrangement of cells within their structure. This exposes students to behavioral robotics, where the focus is on understanding how organisms or entities interact with their environment.

Learning about these interactions builds a strong foundation for studying robotic behavior, control systems, and autonomous decision-making.

3. Swarm Robotics: Swarm robotics involves programming groups of robots to work together, mimicking social animals like ants or bees. When multiple xenobots are released in a shared space, they interact in ways that resemble simple swarm behaviors, such as grouping together or following specific paths.

Students can explore the fundamentals of swarm intelligence, a concept widely applied in areas like search and rescue, environmental monitoring, and space exploration.

4. Sustainable Robotics: Most traditional robots rely on synthetic materials that can contribute to electronic waste. Xenobots, however, are made from biodegradable cells, making them a renewable and environmentally friendly form of robotics.

Learning about sustainable approaches to robotics introduces students to eco-conscious engineering practices and encourages them to consider the environmental impact of technology.

The Impact of Xenobots on STEM Education

Xenobots bring an exciting dimension to STEM (Science, Technology, Engineering, and Mathematics) education. With their interdisciplinary nature, they break down silos between subjects and allow students to learn holistically.

This is especially beneficial for students who might not see themselves as "biology people" or "robotics people"—xenobots appeal to both, showing how fields converge in surprising and meaningful ways.

Additionally, xenobots encourage a more hands-on, inquiry-based learning style, allowing students to formulate hypotheses, design experiments, and observe outcomes in real-time. This aligns well with scientific methods, as students are actively engaging in exploration rather than passively consuming information.

The excitement of working with living, programmable entities brings a level of engagement that can deepen students' understanding and foster curiosity.

Looking to the Future

As xenobot technology advances, it will become increasingly accessible and versatile, allowing educators to incorporate it into a broader range of learning modules. These bio-robots have the potential to shift educational paradigms, demonstrating to students that biology and robotics are not separate fields but two sides of the same scientific coin. By cultivating this multidisciplinary perspective, xenobots may inspire the next generation of scientists and engineers to think creatively about the possibilities at the intersection of life and technology.

Xenobots are more than just biological curiosities—they are educational tools that can spark curiosity, foster interdisciplinary learning, and inspire a new wave of innovation in biology and robotics. In bringing together the living and the mechanical, they present a glimpse of the future that is as thrilling as it is educational.

Advancements in Understanding Cellular Behavior Through Xenobot Research

The field of xenobot research is revolutionizing our understanding of cellular behavior and the potential of bioengineering. Xenobots, which are small, programmable "living robots" created from the stem cells of the African clawed frog Xenopus laevis, represent a new form of biological machine. These tiny structures, assembled and guided by the intrinsic properties of living cells, are neither purely biological nor mechanical. Instead, they inhabit a novel intersection of biology, robotics, and artificial intelligence, allowing scientists to explore cellular behavior in unprecedented ways.

Birth of the Xenobot: A New Approach to Cellular Construction

Xenobots emerged from the fusion of developmental biology and computer science. In 2020, a team of researchers from the University of Vermont and Tufts University developed the first xenobots by reconfiguring frog stem cells to move, sense their environment, and even heal themselves. Unlike traditional robots with metal or plastic parts, xenobots are entirely biological and can spontaneously perform tasks without external hardware.

Scientists relied on a computational model to predict which arrangements of cells might yield specific behaviors, revealing that cellular cooperation and function could be "programmed" in new ways. This was a breakthrough, showing that cells could be organized to perform tasks typically associated with mechanical systems.

Exploring the Fundamentals of Cellular Intelligence

Xenobots have helped to deepen our understanding of what cells can achieve collectively. Each xenobot is created from a clump of undifferentiated cells—meaning cells that have not yet specialized.

In the absence of specific instructions, these cells naturally tend to cluster and form more complex structures.

By arranging and observing these cells, scientists can witness the development of behaviors that might otherwise only be observed in multicellular organisms. Xenobot studies offer a simplified model system to study "cellular intelligence"—the capacity of cells to make decisions, coordinate with each other, and respond to stimuli in ways that seem purposeful.

This raises fundamental questions: Can cells alone exhibit intelligence, and if so, how? How do they communicate with one another, and how do they "decide" what to do?

Understanding these aspects of cellular behavior could have profound implications for developmental biology, regenerative medicine, and our broader understanding of intelligence as a natural phenomenon.

Redefining Self-Healing and Regeneration

One of the most remarkable features of xenobots is their capacity to self-repair. Unlike most traditional robots, which require external repair, xenobots are made of biological cells capable of regeneration. This feature has provided scientists with a model to study natural healing processes at the cellular level.

When a xenobot is cut or damaged, the surrounding cells come together to heal the wound, highlighting the intrinsic ability of cells to repair tissue and maintain structural integrity.

The implications of this are vast. By examining how xenobots self-heal, researchers can better understand the cellular mechanics behind regeneration, which could eventually help humans harness similar capabilities in medical treatments.

For instance, scientists hope to apply insights from xenobot regeneration to develop therapies for tissue injuries, organ damage, and degenerative diseases, where stimulating cells to repair damaged tissue could offer new ways to recover lost function.

Cellular Communication and Collective Decision-Making

Xenobots allow researchers to observe cellular communication and coordination in action. In these living robots, cells must "decide" collectively on how to behave, often without centralized control. For example, cells within a xenobot might coordinate to generate movement by contracting and expanding in sync, demonstrating a level of cooperative behavior usually attributed to more complex organisms.

This process raises important questions about how cells "talk" to each other to achieve a common goal. The study of xenobots has shown that even simple cells, when arranged in specific configurations, can exhibit surprisingly complex group behaviors.

This challenges the traditional notion of centralized control, suggesting instead that even basic cells contain a form of intelligence or awareness that enables them to collaborate and adapt.

Research into how cells achieve collective decision-making could influence areas as diverse as synthetic biology, swarm robotics, and artificial intelligence. By better understanding how biological systems organize themselves, scientists can design more efficient algorithms for robotic swarms and create synthetic tissues that replicate the resilience and adaptability of natural tissues.

Environmental Interactions: Sensing and Responding

Xenobots are not only capable of self-directed movement but also of sensing and reacting to environmental stimuli. They respond to physical obstacles, changes in their environment, and even chemical signals. This has given researchers a unique platform for studying how cellular organisms process sensory information and use it to guide their behavior. Studying xenobots' environmental interactions offers insight into how cells detect changes and respond appropriately, a process vital for survival in all living organisms.

For example, a cell's response to chemical signals plays a critical role in immune response, wound healing, and even in diseases like cancer, where cells may start to ignore typical signals.

Insights gained from xenobot research can contribute to medical applications in which it's crucial to control cellular responses, potentially leading to targeted therapies that help cells recognize and eliminate unhealthy tissue or grow only in desired locations.

Implications for Evolutionary Biology

Xenobots have implications for understanding evolutionary processes at the cellular level. Traditional evolutionary theory suggests that organisms evolve over long periods through genetic changes.

However, xenobot research reveals that cell behavior can adapt quickly to new configurations and tasks without any changes to the underlying DNA. The cells forming xenobots are identical to those in a frog embryo, yet they exhibit different behaviors based solely on their new arrangement and environment.

This challenges the notion that evolution and adaptation are solely genetic processes. Xenobot research implies that evolution might also act at the level of cellular behavior, with cells capable of modifying their function and structure to meet the demands of new situations. This opens up intriguing possibilities for the study of evolutionary biology and how organisms might adapt to new environments on a cellular scale.

Ethical and Future Perspectives

As with any pioneering technology, xenobot research brings up ethical considerations. Because xenobots are living structures capable of movement and self-repair, some have raised questions about their potential for autonomous behavior.

Scientists are careful to ensure that xenobots are confined to laboratory environments and currently lack the ability to reproduce or survive outside controlled conditions.

Looking forward, xenobot research is likely to transform fields ranging from medicine to environmental science. In medicine, xenobots could be used to deliver drugs directly to specific areas in the body, clean up toxic waste at a cellular level, or provide models for testing new treatments.

In environmental science, xenobots might one day be employed to remove microplastics from waterways or repair damaged ecosystems.

Xenobots as a Window into Cellular Mysteries

Xenobot research has opened an exciting window into cellular behavior, highlighting cells' extraordinary ability to work together, adapt, and perform complex tasks without a central nervous system. The potential of xenobots to reveal fundamental insights into cell biology, self-repair, communication, and adaptability makes them a transformative tool for science.

As the journey of xenobot research continues, the discoveries made through these tiny living robots may redefine our understanding of life itself.

Chapter 17: The Xenobot Artistic Revolution

A world where synthetic biology and robotics are blurring boundaries between living and non-living, a new art form is emerging that pushes the limits of what we know about creation and life itself. This movement, which we'll call the Xenobot Artistic Revolution, is unlike any artistic trend before it. It doesn't use canvas, clay, or even digital pixels as its medium. Instead, it crafts using living cells, forming tiny programmable organisms called xenobots—the first living robots that combine science, engineering, and biological aesthetics.

This fusion of art and life opens up questions about the nature of creativity, life, and the potential of technology to interact with organic life on a fundamental level.

The Birth of Xenobots and a New Art Form

Xenobots were first created by assembling embryonic frog cells into tiny, simple organisms that could move, self-heal, and respond to environmental stimuli. These synthetic organisms were developed by scientists and roboticists to explore potential applications for drug delivery, environmental clean-up, and even future healthcare innovations.

But beyond these functional aspects, xenobots brought something entirely unexpected: they became a medium for expression, with life and aesthetics intertwining in ways never before imagined.

The xenobot's organic origins and synthetic control struck artists and thinkers as profoundly symbolic. These creatures represented the blending of life and technology, offering a new kind of creative platform. By programming living cells to form particular shapes, movements, and responses, scientists inadvertently pioneered an art form that could not be replicated in paint or sculpture. The unique medium of xenobots introduces a new set of possibilities and constraints, presenting art not as a static display but as an evolving, dynamic interaction.

Xenobots as Living Sculptures

Unlike traditional sculpture, where an artist molds static materials into permanent forms, xenobots are mutable. They are alive in the sense that they grow, adapt, and sometimes even 'heal' if damaged. Artists working with xenobots explore how shape and movement can interact to create a new type of art.

By modifying the structural design of these organisms, they manipulate the physical properties of life itself.

For example, by arranging the cells of a xenobot to form a particular pattern, artists can influence how the organism moves and interacts with its environment. A xenobot might "dance" in a specific pattern or pulsate to a rhythm determined by its shape.

These "living sculptures" possess a unique aesthetic appeal as they move, transform, and respond to external stimuli. This is not an art piece on display; rather, it is an artwork performing, adapting, and existing in real-time.

Biodesign and Bio-Expression

The artistic manipulation of xenobots draws from principles of biodesign, where design and biology intersect to create innovative, sustainable solutions. In the case of xenobots, biodesign reaches a new level as artists work directly with life's building blocks. Each cellular structure becomes a brushstroke, and each programmed behavior becomes a chosen artistic intention.

Biodesign artists working with xenobots have been inspired by the natural world—patterns seen in plant growth, animal movement, or even the collective behaviors of insect colonies. By programming xenobots to mimic these patterns, artists find ways to represent natural beauty in ways that would be impossible using inanimate materials.

This bio-expression can manifest in fluid, spontaneous movement, or through designs that encourage the xenobots to form complex patterns when observed in a group.

Ethical and Philosophical Reflections

The Xenobot Artistic Revolution also raises profound ethical and philosophical questions. Traditional artists might consider the implications of using non-living materials, but in the case of xenobots, artists are working with life itself. This art form raises questions about the role of humanity in creating and manipulating life.

Can living organisms be considered art if they were created primarily as scientific tools? Where is the line between creation and creation for manipulation? Does this art honor life, or does it merely use life as a means to an aesthetic end?

For many artists and scientists, the answers are deeply personal, and they emphasize a respect for life at every stage of creation. Others argue that the very act of manipulating life for art inherently changes how we view life itself. Art becomes a vehicle for philosophical exploration, one that touches on themes of existence, purpose, and the nature of creativity.

Xenobot Art as a Reflection of Nature and Technology

Xenobot art also serves as a mirror, reflecting humanity's relationship with nature and technology. Historically, art has depicted nature in various forms: landscapes, animals, and abstract representations of life's processes. Xenobots bring this relationship to a new dimension by embodying the tension between natural evolution and technological intervention.

They raise questions about control, unpredictability, and the boundaries of human intervention in nature. Unlike a painted landscape or a sculpted figure, xenobot art is part of nature, behaving in ways that no human can entirely predict.

For instance, by setting xenobots loose in an aquatic environment, artists can observe them forming shapes or moving in patterns that mimic natural behavior but with a hint of controlled randomness. This blending of artificial programming and natural response serves as a poignant metaphor for the way technology interacts with, alters, and sometimes coexists harmoniously with nature.

The Future of Xenobot Art

The future of xenobot art lies in expanding its potential for personal expression, social commentary, and environmental engagement. Imagine a xenobot art installation in which the organisms respond to real-time data about pollution, their movements reflecting shifts in water quality or air purity.

In this case, the art becomes both an expression of life and a form of environmental awareness, revealing the impact of human activity on the natural world.

As scientists gain more control over xenobot behavior, there may also be room for interactive art pieces where viewers can affect how the xenobots behave. With a smartphone or another interface, audiences might interact with xenobot art by changing temperature, light, or sound in its environment, prompting different responses from the organisms.

This form of art would make each viewer a collaborator, creating an even more dynamic and personalized experience.

A New Understanding of Art and Life

The Xenobot Artistic Revolution offers a new language of creativity—one that bridges biology and robotics, art and science, humanity and the organic world. Unlike any other art form, xenobot art invites us to look at life not just as a passive subject but as an active participant in the creative process. It redefines what it means to be an artist, as the process of creation becomes one of guidance rather than control.

In this new frontier, art is not merely something to be observed; it is something alive, constantly evolving, blurring the lines between life and technology, and challenging our most fundamental ideas about creation and existence.

The Xenobot Artistic Revolution reminds us that art, at its core, is about transformation—and that in the age of synthetic biology, transformation may one day be as alive as we are.

Exploring the Intersection of Xenobots and Art

Imagine a creature as small as a speck of dust, made not of metal or plastic but of organic cells—living cells that function together as a miniature, programmable biological machine. These remarkable creations, called xenobots, were first conceived in 2020 through an ambitious collaboration between scientists and roboticists from the University of Vermont and Tufts University. Named after the African clawed frog (Xenopus laevis) from which they draw their cells, xenobots represent an intriguing blend of biology, robotics, and now, even art.

While initially developed for scientific and medical purposes—such as navigating the human body to deliver drugs or clear clogged arteries—xenobots have found an unexpected place in the creative world. By challenging traditional boundaries between science and art, xenobots inspire new artistic visions, pushing forward questions about life, design, and the nature of creation itself.

A Living Medium for Art

What happens when art gains a heartbeat, or when the medium itself can move, interact, and respond? Xenobots invite us to rethink art as a static creation. Traditional art forms—like painting, sculpture, or digital art—are fixed in their final state, where an artist's intent is set and complete. Xenobots, however, introduce the radical concept of "living art"—a piece that's constantly in flux, morphing and adapting based on its environment and biological code.

In creating xenobot art, scientists and artists use programming techniques to guide the movements and interactions of these biological robots. Like cells in the body, xenobots communicate through chemical and mechanical signals, which can be orchestrated to produce intricate, almost dance-like motions. Artists use algorithms to dictate these behaviors, choosing specific patterns and forms that emerge as the xenobots interact with each other and their surroundings.

The results are mesmerizing: swarming motions, coordinated waves, and even simple "social behaviors" that can evoke awe, curiosity, or contemplation.

Art at the Intersection of Biology and Technology

The artistic process behind xenobots requires a blend of both biological understanding and digital technology. Xenobots are designed using computer simulations, where algorithms test various biological configurations to see which designs will be stable and capable of movement.

The result of this design is an organism—one that functions in ways that can be controlled yet remains fundamentally unpredictable, evolving over time.

For artists, this intersection offers unprecedented freedom to experiment with living materials. Unlike virtual simulations or traditional robotics, xenobots are organic, and their behavior can change in ways that reflect real-life biological responses to stimuli.

This means that each piece of xenobot art is, to an extent, unique. Its movement patterns, interactions, and responses will not be identical in any two instances, giving each "performance" a fleeting, unrepeatable quality similar to live theater or dance.

New Perspectives on Bioethics in Art

As xenobots tread this new artistic ground, they raise ethical questions that both scientists and artists must confront. Creating and manipulating living organisms, even at a microscopic scale, forces us to examine our responsibilities as creators.

Is it ethical to treat a living organism as an artistic medium? What implications does this have for future work at the crossroads of life and technology?

For now, xenobots are created using clusters of cells without a nervous system, meaning they cannot feel pain or experience awareness. This alleviates some ethical concerns, as xenobots are not sentient in any traditional sense. Still, as these creations become more sophisticated and as artists and scientists push the boundaries of what's possible, ethical considerations remain a crucial part of the conversation.

Some argue that xenobot art should prioritize sustainability and transparency, ensuring that resources are used wisely and that the creations serve a meaningful purpose, whether it's for aesthetic, scientific, or ecological impact. Others see xenobots as a new mode of expressing complex themes, using their unique attributes to explore ideas around life, agency, and the relationship between humanity and nature.

A Gateway to New Forms of Expression

By merging the organic with the synthetic, xenobot art encourages us to look at life and technology with fresh eyes. Xenobots have the capacity to function as semi-autonomous agents, allowing for a surprising level of improvisation in artistic expression. Artists can design a particular configuration for xenobots and then simply observe as the organisms create their own paths, responding to their environment in real time. This can lead to dynamic, surprising, and often beautiful patterns—organic expressions of art that no human could directly produce.

One xenobot art project, for instance, involved placing hundreds of tiny xenobots in a petri dish, where they formed ever-changing patterns reminiscent of flocking birds or schooling fish. As they moved, interacted, and rearranged themselves in response to subtle environmental cues, they created an ephemeral display, like a painting being continuously reimagined.

An Evolutionary Step in Art and Science Collaboration

Xenobot art also represents a profound evolution in the collaboration between artists and scientists. Traditionally, art and science have been seen as separate, even oppositional fields, but xenobots blur that line entirely. Artists rely on scientific knowledge to understand and utilize the biological mechanisms that make xenobots move, while scientists appreciate the artistic vision that expands their creations beyond practical applications. Together, they craft new forms of expression that neither discipline could achieve on its own.

Such interdisciplinary collaborations encourage both fields to ask fundamental questions: What is life? What does it mean to create? And where is the line between artificial and natural life?

Future Potential and Vision

Looking forward, xenobots could open up endless possibilities for art installations and performances. Imagine an exhibit where visitors watch as xenobots autonomously create living sculptures that dissolve and reform, or a performance where xenobots respond to sounds or lights, creating choreography in response to music. We might even see wearable art made from xenobot clusters, reacting to the movements of the person wearing them.

As xenobots evolve, so too will the opportunities for artistic expression. Perhaps xenobots are the first glimpse of a future in which life itself can be part of art—an art that's not only seen but lived. By intertwining biological and digital worlds, xenobot art serves as a reminder of the unexpected beauty that arises when disciplines intersect, challenging and reshaping our notions of both science and creativity.

Creating Bio-Art Using Living Machines as a Medium

As technology advances, we're constantly redefining what art can be. One of the latest and most intriguing developments in this journey is the use of living machines—specifically xenobots—to create bio-art. This art form combines biological organisms, robotics, and artistic intent, crafting works that are not only alive but can also grow, move, and even evolve over time. Xenobots, a type of microscopic organism engineered from frog cells, are an ideal platform for this emerging field. The possibilities for bio-art using xenobots are vast, sparking both ethical discussions and an artistic renaissance that blurs the boundaries between life and machine

Xenobots are synthetic organisms created by scientists who program biological cells to function in specific ways. The cells, often derived from frog embryos (the African clawed frog, Xenopus laevis), are arranged to form small, living machines that can perform simple tasks. Xenobots are not traditional robots made from metal or plastic; instead, they are clusters of biological cells designed to move in certain ways, carry small payloads, and interact with their environment.

Xenobots can even self-repair and self-replicate under certain conditions, making them both adaptable and resilient.

This adaptability is precisely what makes xenobots so intriguing for bio-art. Unlike conventional robotic installations that require external programming and maintenance, xenobots are self-sustaining systems. Artists who use xenobots as a medium are working with a material that has its own agency and potential for unexpected, organic expression. As living entities, xenobots possess an inherent unpredictability that lends itself well to creative exploration.

The Role of Artists in Xenobot Art

Artists working with xenobots are tasked with shaping life at a cellular level, designing patterns and forms that leverage the inherent behaviors of these organisms. The artist does not control every detail; instead, they design initial conditions and frameworks, allowing xenobots to "perform" in ways that evolve autonomously over time.

This new medium allows artists to think of their work as ecosystems, blurring the line between creation and cultivation.

By designing xenobot bio-art, artists engage with questions that challenge our understanding of life, ethics, and autonomy. What does it mean to create living art that has agency? How do we reconcile the aesthetics of bio-art with the ethical concerns surrounding manipulation of living cells? These questions are integral to the process of creating bio-art using living machines, pushing artists to think deeply about their work beyond its physical form.

Methods of Crafting Bio-Art Using Xenobots

The process of creating bio-art with xenobots begins in the lab, where scientists and artists work together to build the fundamental structures of these organisms. Using stem cells from frog embryos, scientists carefully sculpt cell clusters that are capable of movement, reaction, and basic task completion.

While the programming of xenobots for scientific purposes often focuses on practical applications like environmental cleanup or medical research, the art-oriented programming emphasizes aesthetics, interaction, and transformation.

To give xenobots artistic expression, artists experiment with different cellular arrangements, embedding behaviors that respond dynamically to stimuli such as light, chemicals, or other environmental factors. This interactive quality allows the artwork to "come alive" in a way that traditional media cannot.

For instance, a piece could be designed so that xenobots respond to the proximity of viewers, forming shapes or patterns that change in real-time. This creates a form of art that's not only visually engaging but also immersive and deeply interactive.

Some artists explore the potential of xenobots to replicate and morph, turning bio-art into a form of generative art. By crafting initial designs that encourage self-duplication or mutation, artists create works that evolve over time. The xenobots grow, split, and create new formations, each stage reflecting a new phase of the artist's vision.

This evolving nature of xenobot-based art forces viewers to confront the boundary between creator and creation, as the art seems to take on a life of its own.

The Ethics of Bio-Art and Living Machines

Creating bio-art with xenobots raises profound ethical questions, some of which challenge our current understandings of art and science. At what point does a living machine become a life form worthy of ethical consideration? Is it fair to manipulate life, even at a cellular level, for the purpose of art? These questions have no easy answers, but they're essential to the discourse surrounding bio-art.

Many artists argue that using xenobots as a medium brings attention to these ethical dilemmas, encouraging viewers to think about the implications of bioengineering, artificial life, and the future of technology. In this way, xenobot art is not just about aesthetics but also about sparking critical conversation.

It asks us to reconsider our relationship with nature, our responsibilities toward living organisms, and the ethical limits of scientific and artistic experimentation.

Practical Applications Beyond Art

Though primarily created as a new artistic medium, xenobot art has potential applications in education, medicine, and environmental science. Educators can use bio-art installations to teach students about cellular biology, robotics, and ethics. Xenobot art exhibits allow the public to experience scientific concepts firsthand, making abstract ideas tangible and engaging. In medicine, xenobot art research could provide insights into regenerative medicine and wound healing. The study of self-replicating xenobot structures, for example, may eventually lead to innovations in tissue engineering.

Environmentally, xenobot-based bio-art could inspire methods of bioremediation, where living machines are used to clean up pollutants or restore damaged ecosystems. While traditional robotic solutions are limited by their reliance on metal and electronic components, xenobot art encourages new approaches that blend organic materials with machine functionality.

Future Perspectives

The future of xenobot bio-art is ripe with possibilities. As technology advances, artists will gain finer control over xenobot behaviors, allowing them to create even more intricate and expressive works. We might one day see large-scale xenobot installations that adapt and change over the seasons, mimicking ecosystems that respond to environmental shifts.

Moreover, as ethical frameworks evolve alongside these innovations, society may become more accepting of living machines as an artistic medium. This acceptance would open doors for even more ambitious xenobot art projects that challenge our perceptions of art, science, and the definition of life.

In this way, xenobot bio-art represents a powerful fusion of creativity, technology, and biology. It invites us to explore the boundaries of what we consider life and to reimagine the role of art in a rapidly advancing world. Through the work of these pioneering artists and scientists, xenobot bio-art is set to shape not only the future of art but also the future of humanity's relationship with the living world.

Chapter 18: Challenges in Xenobot Autonomy

In recent years, the emergence of "xenobots"—programmable, microscopic living robots made from frog cells—has opened a new frontier in biocomputing, regenerative medicine, and robotics. Created from skin and heart cells of the African clawed frog (Xenopus laevis), xenobots are capable of movement, sensing, and even self-repair, offering a glimpse into a future where living machines could assist in a range of medical, environmental, and technological tasks.

However, achieving full autonomy for xenobots—where they can operate independently and perform complex tasks without external control—remains a considerable challenge.

1. Energy Sources and Longevity

For xenobots to be fully autonomous, they need sustainable energy sources. Currently, xenobots rely on the naturally occurring energy reserves within their cells, but these reserves are limited. Once the initial energy stores are depleted, xenobots lose their capacity to function effectively.

Ensuring that xenobots can obtain, store, and utilize energy over extended periods, without external intervention, is crucial for enabling them to carry out long-term tasks.

Potential solutions, such as incorporating energy-harvesting molecules or integrating with nutrient sources, are promising but complex. Additionally, scientists are exploring synthetic pathways to "feed" xenobots, perhaps by embedding metabolic capabilities within them.

However, the integration of synthetic biology in a way that does not interfere with their structural or functional integrity is still a challenging and underdeveloped field.

2. Complex Decision-Making

Another significant challenge in xenobot autonomy lies in developing sophisticated decision-making abilities. Currently, xenobots are programmed to follow simple, pre-determined instructions that guide their movement or shape adaptation.

For instance, xenobots can be instructed to move in a specific direction or respond to a light stimulus, but they lack the ability to evaluate complex environmental variables and make adaptive choices.

Creating autonomous xenobots requires more than simple response mechanisms; it requires some level of "intelligence." Researchers are working on developing neural pathways and biological circuits within xenobots that would enable them to respond dynamically to changes in their environment.

The main hurdle here is developing a balance between complexity and efficiency. Incorporating biological circuitry that can process information is difficult in such a tiny and delicate organism, especially when neural-like structures could demand more energy, higher complexity, and potentially reduce the xenobots' longevity.

3. Environmental Adaptability

For xenobots to work autonomously in various settings—whether in the human body or polluted environments—they need to adapt to diverse conditions such as temperature, acidity, salinity, and toxin levels. Natural cells, while adaptable to some extent, are still limited by the physical and biochemical constraints of their origins as frog cells. When placed outside their natural habitat, cells can deteriorate or lose function due to stress.

Scientists are investigating ways to genetically or chemically enhance xenobots to tolerate harsh environments. For example, they could introduce stress-resistant properties through gene editing or molecular additives that increase cell stability. However, these solutions are difficult to balance because they must be implemented without hindering xenobots' other functions or complicating their biological makeup.

4. Size and Scalability

Most xenobots are only a few millimeters wide, and while this tiny size is advantageous for tasks requiring micro-interventions, it limits their functional capacity. Increasing xenobot size could allow for more complex structures and functions, but this comes with challenges. Larger xenobots would demand more energy, have greater structural complexity, and might require intricate control mechanisms to coordinate their activities.

Moreover, scaling up xenobots risks losing the precise cellular configurations needed for their desired functions. Scientists are working on finding ways to preserve the delicate cellular patterns while allowing for more flexibility in size. Alternatively, another approach involves creating "swarms" of smaller xenobots that work together like a colony of ants, thus collectively performing tasks that an individual xenobot cannot accomplish on its own. Coordinating such swarm behavior autonomously, however, requires robust communication systems and decentralized decision-making algorithms.

5. Ethical and Safety Considerations

Autonomous xenobots raise ethical and safety concerns that must be addressed before they can be widely deployed. These concerns revolve around issues of control, containment, and unintended consequences.

For instance, an autonomous xenobot could potentially continue performing a task beyond its intended purpose, creating risks if it were to enter unintended environments or interfere with ecosystems.

Another ethical issue relates to xenobots' potential self-replication. While xenobots currently lack self-replicative abilities, future designs may push toward more advanced biological functions that allow them to self-repair or even reproduce. This ability could lead to ethical dilemmas around the "status" of xenobots, as well as risks related to their containment. Scientists are actively exploring ways to ensure that xenobots remain controllable, such as encoding programmed lifespans into xenobot structures or developing methods for safely deactivating them if necessary.

6. Communicative Abilities

Autonomy in xenobots also requires some form of communication between individual xenobots or with external systems. In many applications, xenobots would need to share information about their environment or their status with other xenobots to work collaboratively on a task.

Communication at such a microscopic scale presents unique challenges, as xenobots lack sophisticated nervous systems or biochemical pathways for transmitting signals across distances.

Researchers are exploring bio-chemical signaling as a potential solution, where xenobots could release specific molecules to trigger responses in nearby xenobots. Another approach might involve engineering xenobots to sense and respond to magnetic or electric fields, thus allowing them to communicate in non-chemical ways.

These methods, however, must be finely tuned, as uncontrolled signals could disrupt xenobot behavior or produce unintended interactions.

7. Self-Healing and Resilience

While xenobots are capable of limited self-repair, improving this ability is essential for long-term autonomy, especially in environments where damage or stress is inevitable. Living cells have natural repair mechanisms, but they are not optimized for the specific structural configurations of xenobots.

Self-healing in xenobots could be enhanced by selectively integrating regenerative cells or by programming xenobots to regroup and reassemble in response to physical damage.

However, the addition of regenerative properties also introduces complexity. Cells specialized for self-repair may not align with the other roles required in a xenobot, such as motion or sensing. Scientists are experimenting with different cell combinations to achieve both resilience and functionality, a process that involves extensive trial and error.

The pursuit of autonomous xenobots involves navigating a complex web of biological, technological, and ethical challenges. Energy sustainability, decision-making, environmental adaptability, size limitations, ethical concerns, communication, and resilience are just some of the obstacles that need to be overcome. Each of these challenges brings xenobot researchers closer to a future where living, programmable organisms can operate independently, potentially transforming fields as diverse as medicine, environmental science, and beyond.

As researchers address these challenges, they are developing foundational knowledge that may one day allow xenobots to function autonomously, safely, and effectively in the world around us.

Limitations and Hurdles in Achieving True Xenobot Autonomy

Xenobots, the world's first living robots created from frog cells, represent a fascinating step forward in biotechnological research. Crafted by repurposing cells from the African clawed frog (Xenopus laevis), these tiny biological machines hold promise for many applications, from environmental cleanup to personalized medicine. Yet despite their immense potential, xenobots face a suite of technical, biological, and ethical challenges that hinder their path toward achieving true autonomy. Below, we explore the limitations and hurdles in realizing fully autonomous xenobots that can operate independently, adapt, and perform complex tasks.

1. Limited Decision-Making Abilities

Xenobots currently lack advanced cognitive mechanisms that would allow them to make complex decisions. They are simple entities with basic programming and can only respond to very limited stimuli. Unlike machines with microprocessors and complex algorithms, xenobots rely on their physical design and cellular behaviors to carry out predefined tasks. For example, researchers have shaped xenobots to move in certain directions by sculpting their cellular structure, but these movements are rudimentary. They follow pre-set patterns without any true understanding or higher-level decision-making, limiting their ability to respond dynamically to unexpected changes in their environment.

2. Challenges in Energy Management

Energy sustainability is another major hurdle for xenobot autonomy. Living cells require nutrients to maintain their function, yet providing a continuous energy source for these tiny biological machines is difficult, especially in environments where external energy sources are scarce. Xenobots made from frog cells may be able to survive for weeks based on stored nutrients or glucose supplements, but this timeframe is too short for many potential applications. Researchers are investigating ways to extend xenobot lifespan through metabolic engineering or symbiotic energy systems, but integrating these solutions in a way that maintains xenobot functionality and autonomy remains challenging.

3. Limited Sensory Capabilities

Current xenobot models lack advanced sensory systems, which restricts their ability to interact meaningfully with their surroundings. In robotics, autonomy often relies on the machine's ability to perceive and interpret complex environmental signals. Xenobots, however, can only respond to simple stimuli, such as pH changes, light exposure, or physical obstacles, in a highly basic manner. For example, they may move in response to light, but they cannot discern between different light intensities or colors, limiting their sensory capacity and adaptability. To achieve full autonomy, xenobots would need an upgrade in sensory and interpretive abilities to recognize a wider range of stimuli and make responsive decisions.

4. Constraints in Self-Repair and Reproduction

A true autonomous biological machine would ideally have self-repair and, in some cases, self-replication abilities to sustain its functions over time. Xenobots show some potential for self-repair; for example, they can close small wounds due to the natural regenerative properties of their cells. However, the self-repair capabilities are limited and do not equate to the level needed for practical applications. Furthermore, while xenobots have shown an ability to produce "offspring" through a novel form of reproduction known as "kinetic replication," this process is rudimentary and not sustainable across multiple generations. Without more advanced self-repair and replication mechanisms, xenobots will struggle to maintain autonomy over extended periods.

5. Ethical and Environmental Concerns

Ethical and ecological considerations play a significant role in xenobot development, especially as scientists strive to create autonomous versions.

The idea of introducing self-sustaining, self-replicating xenobots into the environment raises concerns about unforeseen ecological impacts, such as unintended interference with natural ecosystems or the risk of xenobots proliferating uncontrollably.

Strict regulatory measures and ethical guidelines are required to prevent potential harm, but these constraints limit the scope of xenobot research and the extent to which they can be autonomous.

Balancing the desire for autonomy with safety protocols remains a difficult hurdle.

6. Dependency on External Control

Current xenobots are dependent on external programming and manual design modifications for task assignment, which constrains their autonomy.

To achieve fully independent xenobot systems, researchers need to develop means for in-situ reprogramming or adaptive learning.

This remains a significant challenge, as it would require implementing a form of biological computing within the xenobots, allowing them to autonomously adjust their behavior in response to real-time feedback.

Such capability would be akin to how some bacteria can adjust their gene expression in response to environmental changes.

However, achieving this kind of flexible adaptability in xenobots is challenging because it demands a sophisticated level of cellular programming beyond what is currently possible.

7. Technical Limitations in Biocompatibility and Scalability

Designing xenobots that are both biocompatible and scalable for real-world applications is complex. To perform tasks autonomously in various environments, xenobots must be able to interact harmoniously with their surroundings without causing harm. This is particularly challenging for medical applications, where xenobots need to be non-toxic and biocompatible.

Moreover, scaling xenobot production for widespread use introduces new technical demands in terms of consistency, reliability, and regulatory compliance. Each of these factors limits the feasibility of achieving autonomous xenobots that can operate reliably across diverse contexts.

8. Integrating Learning Mechanisms

For true autonomy, a xenobot would ideally incorporate some form of learning mechanism. Traditional robots can use algorithms and neural networks to learn from experiences and improve their performance over time, but xenobots lack comparable biological "circuitry" to support this kind of adaptive behavior. While research on synthetic biology and cellular computing has shown some promise in creating cell-based logic circuits, embedding these circuits within xenobots in a way that supports learning and adaptation is still beyond current capabilities.

Without a learning mechanism, xenobots are unable to develop adaptive skills, leaving them limited to pre-programmed behaviors.

9. Cellular Longevity and Stability

The lifespan and stability of the cells used in xenobot construction is another major limitation. While synthetic robots can last for years if properly maintained, biological cells are inherently more susceptible to degradation. Xenobots, created from frog cells, are prone to cellular aging and environmental stressors, such as temperature fluctuations or exposure to chemicals. Without engineering solutions that can extend cellular longevity and improve stability, xenobots will struggle to maintain their functionality over time, making long-term autonomy a distant goal.

10. Lack of Complex Communication Abilities

Another key limitation is the xenobots' lack of communication skills, which hinders the potential for collaborative behavior. Many applications would require xenobots to work together in swarms, sharing information to coordinate tasks like locating contaminants in water or delivering drugs to targeted sites in the body.

However, current xenobots do not possess mechanisms to communicate or relay information to one another. Developing a system for inter-xenobot communication, whether chemical, optical, or through some novel bio-signal, is essential for enabling collective behaviors, a crucial component of true autonomy in biological systems.

While xenobots showcase remarkable promise in biomedicine, environmental science, and robotics, true autonomy remains an elusive goal. The path to fully autonomous xenobots is fraught with scientific, technical, and ethical hurdles, ranging from energy management and sensory limitations to complex cellular requirements and regulatory issues. Overcoming these obstacles will require a concerted, interdisciplinary effort that brings together expertise in synthetic biology, artificial intelligence, and ethics.

Until then, xenobots will remain a groundbreaking yet limited tool in our exploration of living technologies, their autonomy constrained by the very biological systems that make them so unique.

Overcoming Obstacles to Create More Independent Xenobots

The journey toward developing xenobots — programmable, self-assembling biological entities made from frog stem cells — is marked by exciting potential and formidable challenges. The notion of creating independent xenobots capable of carrying out complex tasks with minimal intervention sparks enthusiasm but requires innovative solutions.

Overcoming obstacles to make xenobots more autonomous calls for advances across multiple areas, including control mechanisms, cellular longevity, and ethical considerations.

1. Understanding the Fundamental Control of Xenobots

Xenobots are currently designed with limited programmability, primarily influenced by their initial structure and composition. In the absence of a nervous system, traditional methods of controlling movement and decision-making fall short. Xenobot activity largely depends on inherent cellular properties, with some shapes promoting crawling or swimming while others facilitate specific tasks, such as carrying objects.

To create truly independent xenobots, researchers are exploring techniques to enhance xenobot control. One approach is integrating genetic engineering methods to add programmable cellular responses.

For example, researchers could implant genetic circuits allowing xenobots to react to certain chemical signals or environmental triggers. These genetically integrated pathways would enable xenobots to make basic "decisions" — choosing paths or selecting tasks based on specific environmental cues.

Another promising area is the development of micro-scale "brains" within xenobots. By embedding synthetic neural-like circuits, scientists can potentially introduce low-level decision-making capabilities.

Such neural networks would grant xenobots rudimentary forms of memory, enabling them to alter behaviors based on past interactions with their environment.

2. Addressing Cellular Longevity and Structural Stability

One significant obstacle in xenobot research is maintaining cellular longevity. Since xenobots are made from living cells, they are subject to the same limitations as any other biological material. Over time, xenobot cells experience wear and degradation, reducing their functionality and lifespan.

Without longer-lasting cells, the xenobots' ability to carry out sustained tasks remains limited, as their lifespans are short — often no more than a few weeks.

Researchers are tackling this issue by experimenting with modifications at the cellular level. Some are looking into gene editing to increase cellular resilience, making cells more resistant to natural decay or damage. For instance, researchers might engineer cells to express proteins associated with stress resistance or cellular repair, potentially extending xenobot viability.

A parallel approach is the integration of biomaterials or synthetic components that can offer structural support. By combining biological cells with biocompatible polymers or hydrogel materials, scientists hope to strengthen xenobots without compromising their biological functions. These hybrid structures could prevent premature cellular breakdown and help sustain xenobot forms over longer periods, opening up possibilities for extended tasks or exploration in complex environments.

3. Navigating Energy Constraints

A critical aspect of independence is energy, and current xenobots lack mechanisms to autonomously source or renew their energy. Xenobots presently rely on the inherent metabolic resources of their cells, which are depleted quickly. For extended functionality, xenobots would need a method to continually generate or absorb energy.

One avenue researchers are exploring is the addition of energy-harnessing components within xenobots, potentially through photosynthetic or chemical metabolic pathways.

This would enable xenobots to convert sunlight or absorb surrounding nutrients into usable energy. Using plant-derived cells capable of photosynthesis, for example, could allow xenobots to survive longer in light-rich environments, while the incorporation of micro-scale fuel cells could harness chemical energy from their surroundings.

Efforts to miniaturize biofuel cells are especially promising, as these devices could offer xenobots a way to convert organic materials into energy autonomously. By integrating biofuel cell technology, xenobots could, in theory, "feed" on small organic particles in their environment, providing a sustainable power source for independent activity.

4. Programming Adaptive Behaviors in Dynamic Environments

Autonomous functionality necessitates the ability to adapt to changing conditions, which remains a challenging frontier for xenobot research. Current xenobots operate within pre-defined parameters; they are often designed to perform a single task in a stable environment.

To create xenobots capable of acting in unpredictable surroundings, researchers need to instill adaptive behaviors — an area that overlaps with artificial intelligence (AI) and machine learning.

One promising strategy is to use adaptive programming techniques based on reinforcement learning. By embedding simple reinforcement learning algorithms into xenobot control systems, scientists could enable them to adjust their behaviors in response to environmental stimuli.

For instance, a xenobot with reinforcement learning capabilities might learn to avoid obstacles or alter its course when confronted with physical barriers.

A complementary approach is designing xenobots with modular behaviors. In this approach, xenobots would possess a "toolbox" of different behaviors, and a basic algorithm would determine which action is appropriate based on the situation.

Over time, xenobots could use feedback from their environment to select and refine the optimal behaviors, much like an autonomous robot adjusting its movements in real time.

5. Ensuring Safety and Ethical Considerations

As xenobot independence grows, so too must ethical considerations. Autonomous xenobots, especially those with energy-sustaining abilities or adaptive behaviors, raise important questions about control, unintended consequences, and the potential for ecological impact.

For example, xenobots with the ability to self-repair or renew energy might inadvertently persist in environments where they are not intended, disrupting local ecosystems.

To address these ethical concerns, researchers are designing xenobots with built-in "failsafe" mechanisms. These could include programmed lifespans, making xenobots cease functioning after a certain period, or trigger-based self-destruction mechanisms if they enter a designated area.

Moreover, strict regulatory frameworks will likely be required to govern xenobot design, ensuring that only safe and environmentally neutral designs are deployed outside of laboratories.

Many are also advocating for transparency in xenobot research, suggesting open databases and data-sharing platforms for researchers to report on xenobot development, behavior, and testing outcomes. Through these measures, researchers aim to create an ethical, controlled pathway toward more autonomous xenobots while addressing public and scientific concerns.

6. Looking Forward: The Future of Independent Xenobots

The vision of fully autonomous xenobots is both fascinating and complex, requiring multidisciplinary collaboration among biologists, engineers, and computer scientists. By advancing cellular resilience, developing programmable control systems, and tackling ethical concerns, the field edges closer to creating xenobots that could one day operate independently.

Such xenobots could perform tasks in medical settings, environmental monitoring, or even disaster relief, where they might navigate hazardous areas or monitor pollutants autonomously.

Though challenges remain, researchers are making steady progress in overcoming the obstacles to creating more independent xenobots. By pushing the boundaries of bioengineering, we may soon reach a point where xenobots represent a bridge between biological organisms and autonomous machines, a remarkable testament to human innovation in the biological sciences.

The journey may be long, but each breakthrough brings xenobots one step closer to becoming an independent and functional part of our technological landscape.

Chapter 19: Xenobots and Space Exploration

As humanity sets its sights on distant planets, moons, and beyond, the challenges of space exploration grow increasingly complex. Traditional technologies and methods, though robust, face limitations in the unforgiving environments of outer space.

Enter xenobots – tiny, programmable organisms made from frog cells that have recently captivated scientific and public attention.

These living machines offer unique capabilities, such as self-healing, adaptability, and environmental responsiveness, which could make them indispensable tools in advancing space exploration.

Xenobots are not robots in the traditional sense, nor are they purely biological organisms.

Developed from the stem cells of the African clawed frog (Xenopus laevis), these millimeter-sized organisms are designed using computer algorithms and assembled into specific configurations to perform simple tasks, like moving towards a target or carrying small objects.

Unlike conventional machines, xenobots are soft, flexible, and capable of self-repair, meaning they can potentially adapt to a variety of environments – a quality invaluable in space, where conventional machines often suffer under extreme conditions.

By combining the latest advances in computer modeling, bioengineering, and developmental biology, scientists are exploring the full potential of xenobots as self-replicating, biodegradable, and highly adaptable entities.

This makes them prime candidates for applications in space, where autonomy, resilience, and efficiency are paramount.

The Unique Challenges of Space Exploration

Space exploration demands tools and technologies capable of withstanding extreme temperatures, radiation, and gravitational shifts. In addition, the remoteness of space travel makes maintenance difficult or impossible.

Rovers, landers, and satellites serve as essential equipment, yet they face rapid degradation in harsh environments, from the scorching heat on Mercury to the icy landscapes of Europa. Xenobots, with their unique self-healing properties and adaptive functions, may overcome many of these limitations.

Key obstacles for space-bound xenobots include radiation resistance, survival in varied atmospheric conditions, and enduring long-term missions.

However, because xenobots are living, they possess an innate ability to adapt and repair themselves, qualities that make them ideal for tasks where traditional machinery might fail.

Potential Applications of Xenobots in Space Exploration

1. Microbial Cleanup and Bio-Remediation: On long-term space missions or other planetary surfaces, contamination from Earth-based microbes or pollutants presents a critical challenge. Xenobots could potentially be deployed to consume and break down waste materials, including harmful microbes that may contaminate research environments. Due to their biodegradability, xenobots could self-degrade after completing their tasks, leaving minimal impact on the environment.

2. Exploring Subsurface Environments: Some of the most intriguing environments for life on other planets lie beneath the surface, such as the potential subsurface oceans on Jupiter's moon Europa or Saturn's moon Enceladus.

Sending xenobots equipped with sensory and movement abilities into these environments could allow them to penetrate ice layers and explore aqueous environments in ways that rigid machinery cannot.

Xenobots could potentially navigate narrow crevices and icy waters, reporting back data and even capturing tiny samples for analysis. Due to their small size and flexibility, they could reach areas that larger machines cannot, opening new possibilities for planetary exploration and the search for extraterrestrial life.

3. Self-Healing Spaceships and Infrastructure: Self-repair is a primary advantage of xenobots over traditional machinery. Engineers envision spacecraft that could include embedded xenobots capable of repairing microcracks or shielding components from environmental stress.

For example, if xenobots could detect a crack in a space module, they might activate and travel to the location to seal the breach, reducing the need for costly and complex repair missions.

This property could be extended to permanent human habitats on the Moon or Mars, where xenobots could patrol structures, identify weaknesses, and initiate minor repairs.

Such a design could significantly extend the lifespan of space infrastructure and enhance safety for astronauts.

4. Medical and Biological Support for Astronauts: Space environments pose risks to human health, including muscle atrophy, radiation exposure, and bone density loss. Xenobots could play a role in long-term health monitoring and treatment by being programmed to deliver medicines or monitor biomarkers in real time. With customization, xenobots might one day be used within the human body to address specific health challenges, such as boosting immune responses against pathogens or repairing cellular damage caused by cosmic rays.

5. Terraforming and Bioprinting on Planetary Surfaces: Terraforming – modifying the environment of a planet to support Earth-like life – is still in the realm of science fiction. However, xenobots could serve as the first step in preparing planetary surfaces.

By interacting with soil, moisture, and organic material, they might help to cultivate basic bio-organic materials on barren surfaces, slowly transforming them into more habitable environments.

The adaptability of xenobots could be further harnessed in 3D bioprinting applications, producing small structures or even functional biological tissues on-site.

This capability would significantly reduce the amount of materials needed to be transported from Earth, enhancing the feasibility of sustained exploration missions.

Challenges and Limitations

While the potential applications of xenobots are compelling, significant challenges remain. Firstly, radiation in space could disrupt the xenobots' cellular structures, impacting their function. While their natural resilience offers some protection, researchers would need to genetically or chemically enhance xenobots for prolonged missions.

Secondly, the question of control and predictability must be addressed. Living organisms, even those as programmable as xenobots, are influenced by their environment, which means that strict control over their behavior may be challenging.

Finally, ethical and regulatory concerns around deploying living organisms in space must be carefully considered. As organisms that could potentially replicate under certain conditions, xenobots raise questions about containment and environmental impact.

Robust ethical guidelines and international regulations would be essential for managing these concerns responsibly.

The Future of Xenobots in Space

Looking forward, xenobots could become an integral part of the space exploration toolkit. With further development, they could support humanity's most ambitious goals, from colonizing Mars to exploring distant moons.

The adaptability, self-repair, and minimal environmental impact of xenobots make them highly appealing for future missions that demand ingenuity and resilience.

As the technology evolves, scientists envision creating more complex xenobots capable of performing a broader range of tasks, from microscopic analyses of alien soil to autonomous repairs on space equipment. With each breakthrough, xenobots bring us closer to overcoming the many limitations of space travel and extending humanity's reach into the cosmos. In a future where biological and technological boundaries increasingly blur, xenobots may stand as a testament to humanity's ability to innovate in the face of the unknown.

The potential for xenobots in space exploration is vast. These tiny, adaptable organisms have the potential to revolutionize how we explore, build, and survive in space.

Xenobots in Space Exploration: A New Frontier

As humans venture farther into the cosmos, the technologies that support space exploration need to evolve rapidly to withstand the unique and demanding conditions of outer space. Traditional robotics and artificial constructs, while powerful, often face limitations in extreme environments. Xenobots—biologically engineered, programmable living organisms—offer a new and highly promising alternative for space missions.

By leveraging the adaptability, regenerative capabilities, and efficient energy usage of living cells, xenobots might one day play a critical role in exploring and navigating the vast unknowns of space.

The Basics of Xenobot Design

Xenobots, derived from the stem cells of Xenopus laevis (African clawed frog), are essentially biological robots. These tiny living constructs, assembled through a combination of cellular biology and computer-aided design, can move, respond to stimuli, and even self-repair under certain conditions. Unlike traditional metal-based robots, xenobots are entirely organic, making them naturally biodegradable and capable of regenerating damaged cells.

This unique trait is particularly advantageous for the hostile and unpredictable environment of space.

Xenobots are designed using computer algorithms to model specific shapes and cellular formations, which are then "built" by assembling living cells. These cells communicate and self-organize, creating xenobots that can execute various programmed tasks.

By harnessing the natural properties of cells, xenobots operate with an efficiency that mechanical robots cannot easily match. In space missions, this could mean a reduced need for fuel, better resilience, and lower weight—all essential factors for long-term space travel.

Resilience to Extreme Environments
One of the greatest challenges of space exploration is the extreme environment of space itself. Temperatures swing between boiling hot and freezing cold, and radiation levels are many times higher than on Earth. Traditional materials like metals and plastics deteriorate under these harsh conditions, which compromises their effectiveness.

Xenobots, on the other hand, possess the cellular resilience that allows biological systems to adapt to a variety of environments. They can be modified to withstand different stressors, including temperature variations and certain levels of radiation.

This adaptability is especially promising for tasks in uncharted regions of space, where robots might need to function without frequent repair or human intervention. Unlike traditional machines, which become damaged and lose functionality, xenobots could be programmed to survive, repair themselves, and continue their mission autonomously.

Microgravity Operations and Self-Replication

In space, microgravity can present unexpected challenges for traditional robotic systems, impacting movement and the use of resources. Xenobots, being soft-bodied and responsive to their environment, can navigate in microgravity without complex propulsion systems.

This trait is particularly advantageous in asteroid exploration or in the orbital environment around planets and moons, where xenobots could collect samples, monitor atmospheric conditions, or explore difficult-to-reach locations.

Furthermore, the self-replicating potential of xenobots adds another layer of utility. While not capable of reproduction in the traditional biological sense, they can divide and "assemble" new xenobots under specific conditions. This ability could enable them to grow their population in a space habitat, allowing them to expand their range of operations autonomously.

For example, if tasked with cleaning space debris or assembling small structures in orbit, xenobots could theoretically increase their numbers as needed, thereby reducing the need for resupply missions from Earth.

Advanced Resource Collection and Recycling

Space missions are constrained by limited resources, and every gram of material sent into space is costly. Xenobots offer an innovative solution for in-situ resource utilization (ISRU), which is essential for missions to distant locations, like Mars or the moons of Jupiter and Saturn.

Xenobots could be engineered to gather materials from the environment, such as lunar or Martian soil, and perform tasks like filtering water or breaking down hazardous compounds.

In addition to collecting resources, xenobots can facilitate recycling processes, turning waste into reusable material. Their cellular nature enables them to break down organic waste into base components, which could then be repurposed to support life in space habitats.

This ability to recycle and regenerate resources could drastically extend the duration of space missions and reduce dependency on Earth-based supplies.

Xenobot Swarms for Scouting and Terrain Analysis

In unknown landscapes, such as the surface of Mars or the icy crust of Europa, scouting terrain and identifying hazards are critical. Traditional robotic scouts are usually large and designed to endure harsh terrains, but they are limited in agility and often unable to reach certain areas. Xenobots, being microscopic or small in scale, could operate as swarms, moving collectively to cover large areas, map terrain, and avoid obstacles in real time.

Their natural flexibility and responsiveness allow xenobot swarms to adapt to rugged landscapes, navigate narrow crevices, or enter areas that would be inaccessible for standard robots.

By working together, xenobot swarms can send back detailed data, including topographic maps, temperature variations, and even signs of microbial life. This would not only improve mission safety but also deepen our understanding of these distant worlds.

Biocompatibility and Terraforming Potential
A long-term vision for xenobot use in space lies in their biocompatibility and potential applications for terraforming.

While this concept is still in its infancy, researchers believe that xenobots could be an essential tool in preparing extraterrestrial environments for human colonization.

For instance, they could be programmed to process soil on Mars, turning it into nutrient-rich ground suitable for plant growth.

Their biological composition also makes xenobots more compatible with ecosystems than metal-based robots, reducing the likelihood of contaminating alien biospheres with foreign materials.

Over time, xenobots could even be used to introduce microorganisms to specific environments, initiating a controlled and gradual transformation of alien landscapes.

Challenges and Future Directions

While the potential applications of xenobots in space are staggering, significant challenges remain. The development of xenobots capable of withstanding the full spectrum of space conditions—such as high radiation, vacuum pressure, and extreme temperatures—requires further research.

Moreover, ethical questions arise concerning the introduction of Earth-based life forms into extraterrestrial ecosystems.

Additionally, researchers need to ensure that xenobot replication is carefully managed to prevent uncontrolled proliferation in sensitive environments.

Nevertheless, the future of xenobots in space is incredibly promising. By leveraging the adaptability and resourcefulness of biological systems, xenobots could open up new pathways for exploration, resource management, and even settlement on distant worlds.

As this technology continues to advance, xenobots may well become humanity's most versatile allies in the journey into the stars.

Xenobots as Adaptable Tools for Extraterrestrial Exploration
In the realm of scientific discovery, xenobots represent a unique frontier, an intersection of biology and robotics capable of pushing the boundaries of exploration far beyond Earth's atmosphere.

As we aspire to explore celestial bodies within our solar system and even beyond, xenobots offer a pioneering avenue to carry out complex tasks in conditions where conventional robots or human presence may fall short.

Unlike traditional machines, xenobots—living, programmable entities created from frog stem cells—possess unique attributes that make them adaptable to extreme environments and potentially transformative in space exploration.

Xenobots are synthetic life forms developed from the stem cells of Xenopus laevis, the African clawed frog.

Through precise manipulation, scientists can design these cells to form new, self-organizing structures that function as living robots.

While still small in scale, xenobots are endowed with simple programming that enables them to complete specific tasks, from moving through water and carrying payloads to clustering together in "swarms."

They can heal minor damage and live for several weeks without the need for additional energy sources, traits that open up fascinating applications for space exploration.

Why Use Xenobots in Space?

The environments found in extraterrestrial terrains pose challenges that would easily overwhelm traditional machinery. Factors like extreme temperatures, intense radiation, limited accessibility, and unknown biological hazards require exploration tools that can withstand and adapt to hostile conditions. In this regard, xenobots offer a remarkable degree of versatility. As organic machines, they are inherently capable of responding to environmental stimuli.

They have a self-sustaining biological composition that could adapt to alien terrains, interact with unknown materials, and potentially repair themselves—an invaluable feature for extended missions on remote planets or moons.

Traditional robotic missions rely on sophisticated equipment that, once damaged, is often rendered unusable. Xenobots, however, could regenerate to an extent, potentially enduring damage from cosmic rays or minor physical disturbances. They could be tailored to detect and process environmental changes on other planets, offering a dynamic alternative to rigid, non-biological machines.

Adaptability to Varied Environments

Xenobots' unique adaptability is rooted in their living, organic structure. This makes them well-suited for unpredictable environments. They could survive in freezing temperatures, resist radiation to an extent, and respond to physical obstacles. Their natural structure provides a certain resilience not found in metal-based robots, which often rely on protective layers that are susceptible to cracking, warping, or failing under extraterrestrial conditions.

On a Martian exploration mission, for instance, xenobots could be released to gather soil samples, explore crevices, or analyze microenvironments without the risk of being destroyed by the planet's thin atmosphere, dust storms, or fluctuating temperatures.

This self-organizing capability means they can cover large areas and conduct tasks even when communication from Earth is delayed.

Efficient Exploration through Self-Replication and Swarming

One of the most intriguing aspects of xenobots is their ability to cluster, or "swarm," to perform collective actions. For space exploration, a swarm of xenobots could cover a vast area, collecting critical data from multiple points simultaneously.

This swarming ability is especially beneficial on planetary surfaces like Mars, where vast expanses need mapping or sample collection.

Beyond swarming, xenobots have also shown the preliminary ability to self-replicate under certain conditions. This feature could be revolutionary for space missions by reducing the need for large, single-point units to conduct operations.

In theory, if scientists could harness this controlled self-replication, a small number of xenobots sent on a mission could expand their numbers and create additional "workers" autonomously.

In the distant future, this ability might be leveraged on asteroid or moon missions to continuously reproduce specialized xenobots suited for different tasks, such as mineral extraction, environmental monitoring, and other planetary analyses.

Real-Time Monitoring and On-Site Repair

Xenobots' living composition provides an additional layer of functionality that typical robots lack: real-time responsiveness to environmental changes.

Unlike mechanical probes that must constantly transmit data back to Earth, xenobots could offer a more autonomous approach, enabling preliminary on-site responses to environmental changes.

Their structure enables them to "feel" environmental shifts, which could be harnessed to indicate specific types of chemical or radiation exposures.

For instance, xenobots on a lunar mission might be designed to respond to high radiation zones by clustering and forming protective barriers around sensitive areas of interest, such as ice deposits or specific mineral sources.

This autonomous adaptability would make them particularly effective for exploring unknown areas, as they could initiate protective measures or recalibrate to maximize their functionality based on immediate conditions.

The potential for xenobots to self-heal or repair structural damage is another advantage. In environments like Mars, where dust storms or physical disturbances could cause damage, xenobots with self-repair capabilities could sustain themselves longer and ensure mission continuity.

Rather than sending human teams or traditional machines for costly maintenance missions, self-repairing xenobots would continue to operate autonomously, greatly enhancing mission efficiency.

Biocompatibility and Planetary Protection

Space missions are required to follow strict protocols to prevent contamination of other planets with Earth-based organisms.

As organic entities, xenobots present an ethical and practical solution to this issue.

Scientists could engineer xenobots to naturally degrade after fulfilling their purpose, minimizing environmental risks and ensuring compliance with planetary protection standards.

Additionally, xenobots might be programmed to respond selectively to certain environmental factors, allowing for limited interaction with extraterrestrial soil, water, or even biological material if encountered.

Unlike non-degradable plastics or metal that might pollute alien ecosystems, xenobots, once engineered for biodegradability, would effectively "die off" after their mission, decomposing without leaving harmful traces.

This degradation not only protects alien environments but also reduces the footprint of our explorations, making xenobot missions more sustainable.

A Vision for the Future of Space Exploration

The applications for xenobots in extraterrestrial exploration are as vast as the universe itself. As xenobot technology evolves, researchers envision more sophisticated versions capable of navigating through liquid oceans on Europa, probing the icy surface of Enceladus, or exploring lava tubes on Mars.

Their flexibility in programming and biological foundation means they could be tailored to various exploratory needs, paving the way for groundbreaking research and discovery.

As humanity looks toward interplanetary exploration, xenobots represent an adaptable, efficient, and ecologically conscious means of reaching beyond our world.

Through swarming, self-repairing, and biodegrading capabilities, they could tackle the unique demands of space, offering an entirely new way to observe, collect, and analyze data.

In an era where the challenges of space are still largely unknown, xenobots offer an exciting glimpse of a future where exploration and biology work in tandem to unlock the mysteries of our universe.

Chapter 20: Xenobots and Nanotechnology

In the evolving field of synthetic biology, few topics are as exciting or revolutionary as the merging of xenobots and nanotechnology. Xenobots—tiny biological robots engineered from living cells—already show enormous potential for performing complex tasks on a micro scale. When combined with advancements in nanotechnology, their abilities could extend far beyond what either technology could accomplish alone.

This chapter explores how nanotechnology may augment xenobots, enabling them to perform tasks with remarkable precision in environments that are otherwise inaccessible to traditional robotics or human intervention.

Xenobots, derived from African clawed frog cells (Xenopus laevis), are programmable, self-assembling biological organisms created through advanced bioengineering.

Unlike conventional robots made from metals or plastics, xenobots are composed of living tissue. They are capable of unique biological functions: they can move, self-heal, and respond to their environment autonomously.

Currently, xenobots have demonstrated a capacity for simple tasks such as navigating through fluid environments, pushing tiny objects, and even working together as a swarm.

These abilities mark xenobots as a new class of biological machines, distinctly different from traditional organisms due to their customized design and cellular programming.

However, xenobots currently face limitations, particularly in tasks requiring high precision or interactions with nanoscale elements. This is where nanotechnology, the science of manipulating matter on an atomic and molecular scale, offers transformative possibilities.

Understanding Nanotechnology: The World at the Nanoscale

Nanotechnology operates on the scale of nanometers, one-billionth of a meter. At this scale, materials often display different properties compared to their macroscale counterparts, due to the quantum and surface area effects.

Nanotechnology allows scientists to create materials and devices with highly specific structures, enabling them to interact in complex ways with biological systems. Some of the key advancements include nanoparticles, nanofibers, nanosensors, and nanoscale actuators, all of which are relevant to the development of advanced xenobot capabilities.

By integrating xenobots with nanoscale technologies, researchers can potentially engineer microscopic tools with unprecedented precision, ideal for applications such as targeted drug delivery, environmental cleanup, and even deep-space exploration.

The Potential of Nanotechnology-Enhanced Xenobots

1. Precision in Medicine and Targeted Drug Delivery

One of the most promising applications of nanotechnology in xenobots lies in medicine. Nanotechnology could enable xenobots to deliver drugs directly to specific cells or tissues within the body.

Unlike traditional drug delivery methods, which often result in the drug diffusing throughout the body and causing side effects, nanotechnology-enhanced xenobots could carry nanoparticles loaded with medication.

Once programmed, these xenobots could transport drugs to precise locations, releasing them in a controlled manner directly where they are needed.

Additionally, xenobots enhanced with nanosensors could detect molecular markers associated with disease, such as cancer cells or harmful bacteria, allowing them to respond to early warning signs and initiate treatment or alert medical professionals.

2. Enhanced Environmental Cleanup

Nanotechnology offers innovative solutions for environmental challenges, particularly through the use of nanomaterials that can absorb or neutralize toxic substances. When applied to xenobots, this could lead to effective microcleaning units capable of identifying, capturing, and degrading pollutants on a molecular level.

For instance, nanomaterials with high reactivity could be embedded within xenobots to break down oil spills, microplastics, or even hazardous chemicals in water sources.

Such applications are not only efficient but potentially sustainable, as xenobots are biodegradable and do not contribute to additional pollution when their task is complete. By making them function at a nanoscale, xenobots could detect and eradicate contaminants that would otherwise evade traditional cleaning methods.

3. Nanoscale Communication and Coordination

Xenobots, especially those operating in swarms, could benefit significantly from nanoscale communication systems. Through embedded nanosensors and transmitters, xenobots could communicate with each other using molecular signals, forming complex, coordinated responses to their environment.

This would allow a group of xenobots to distribute themselves intelligently across a region, adapting to new situations and obstacles without human intervention.

In medical applications, for example, a swarm of xenobots could distribute themselves throughout a tissue to search for cancerous cells or deliver targeted treatments.

In environmental settings, they could work together to detect pollution hotspots, optimizing cleanup efforts. With nanotechnology, this type of autonomous, collective behavior becomes far more precise and effective, expanding the potential impact of xenobot swarms in various fields.

4. Self-Repair and Adaptive Functionality

Self-repair is a critical feature for xenobots, but integrating nanotechnology could enhance their durability even further.

For example, nanoparticles could be used to supply xenobots with additional cellular components or energy sources, allowing them to maintain function over extended periods.

Nanoscale structures might even serve as "patches" or repair mechanisms for xenobots that experience cellular damage.

With the inclusion of nanotechnology, xenobots could become adaptable machines that are not only resilient but capable of evolving in response to environmental demands.

5. Integration with AI and Machine Learning

The combination of xenobots, nanotechnology, and artificial intelligence (AI) could enable entirely new possibilities.

With nanotechnology, AI algorithms could be embedded within xenobots on a molecular level, allowing them to process information and make complex decisions in real time.

This fusion would be particularly powerful in applications like medical diagnostics, where xenobots could use machine learning to differentiate between healthy and unhealthy cells. AI could also enable adaptive behaviors, allowing xenobots to respond more intelligently to changing environments.

In practice, AI-enhanced xenobots might use nanotechnology to detect and interpret chemical signals, guiding them to areas of disease within the body or adjusting their cleaning methods depending on pollution types.

This integration of AI and nanotechnology into xenobots could make them some of the most intelligent and autonomous biological machines ever created.

Challenges and Ethical Considerations

While the integration of nanotechnology and xenobots presents exciting prospects, it also raises challenges and ethical considerations. One major concern is the potential for unintended environmental impact. For example, should a group of xenobots malfunction, they could potentially spread beyond their intended target areas, causing unintended biological effects.

Additionally, the use of xenobots with self-repair and adaptive capabilities may raise questions about their autonomy and the potential for unanticipated behaviors, especially in complex ecosystems or human bodies.

Regulatory frameworks will be essential to ensure safe deployment, ethical oversight, and environmental monitoring of nanotechnology-enhanced xenobots. Clear guidelines on their design, testing, and usage will be needed to balance innovation with responsible use. Ensuring transparency and thorough research into the long-term effects of these bio-nanotechnologies will be vital to their sustainable and safe implementation.

The Future of Xenobots and Nanotechnology

The convergence of xenobots and nanotechnology stands at the frontier of scientific innovation, promising unprecedented capabilities in medicine, environmental management, and beyond. This symbiotic relationship between biological robotics and nanoscience is poised to redefine what is possible on both the microscopic and macroscopic scales.

As research progresses, the potential for xenobots augmented with nanotechnology will likely grow, unlocking new applications and answering fundamental questions about life, technology, and their intersections.

The journey into this microcosmic frontier has only just begun, but it promises a future where biology and technology coexist in ways we are only beginning to imagine.

Merging Xenobots with Nanotechnology for Novel Applications

The fusion of xenobots with nanotechnology opens an exciting frontier in science, with the potential to tackle complex problems across medicine, environmental science, and engineering. Xenobots, biological robots developed from frog cells, have demonstrated unique behaviors, such as self-organization, movement, and collective coordination.

When merged with nanoscale technology, these programmable living robots can interact with their environment on an unprecedented level, offering innovative solutions that bridge biology and technology.

Understanding the Basics:
What Are Xenobots and Nanotechnology?

To appreciate the synergy between xenobots and nanotechnology, it helps to understand each component. Xenobots are synthetic organisms built from living cells, typically skin and heart cells from the Xenopus laevis frog, hence the term "xenobot."

Using computer algorithms, scientists configure these cells into tiny structures that can perform specific tasks, like moving toward a target or transporting small objects.

While xenobots are millimeters in size, they have remarkable biological characteristics such as self-repair and the ability to respond to environmental signals.

Nanotechnology, on the other hand, manipulates materials at the molecular and atomic levels, within the range of 1 to 100 nanometers. This field is responsible for creating nanoscale devices that can detect, measure, and interact with matter in highly controlled ways. With capabilities to build materials from the ground up, nanotechnology has already revolutionized fields like electronics, medicine, and materials science.

Integrating these nanoscale functionalities with xenobots opens new avenues for scientific discovery and application.

Synergies in Merging Xenobots and Nanotechnology

Xenobots possess inherent biological features that make them appealing for combining with nanotechnology. They are soft, flexible, and biodegradable, making them ideal for applications within the body or environmentally sensitive areas.

Nanotechnology, with its precision, can enhance xenobot capabilities by enabling them to perform tasks at finer scales or interact with cellular and molecular components in ways that purely biological systems cannot.

For instance, xenobots equipped with nanosensors could detect and respond to specific chemicals, bacteria, or temperature changes. This combination can lead to smart biological devices that can sense, process, and act on environmental data, mimicking some functions of complex organisms. Merging nanotechnology with xenobots thus creates a system that operates with both biological adaptability and nanoscale precision—a powerful tool for tackling challenges that require both.

Novel Applications in Medicine

The medical field stands to gain significantly from xenobot-nanotechnology hybrids, especially in targeted drug delivery, tissue repair, and diagnostics. Imagine a xenobot with nanoscale drug-carrying capsules embedded within its structure. These xenobots could navigate through the bloodstream to reach precise locations in the body, releasing their therapeutic payload only in the presence of specific signals, such as the chemical markers of a tumor. By doing so, these systems could deliver drugs directly to diseased cells, minimizing side effects on healthy tissue—a limitation in traditional drug delivery methods.

Xenobots with embedded nanosensors could also monitor internal health indicators, such as pH, oxygen levels, or the presence of pathogens. In cases of infection or injury, they could release drugs or anti-inflammatory compounds, providing real-time, localized treatment that adjusts dynamically based on the condition of the patient. This level of precision and responsiveness could be groundbreaking for personalized medicine, where treatments are tailored not only to the individual but also to the specific needs of their cells and tissues at any given moment.

Furthermore, xenobot-nanotech hybrids might one day play a role in tissue engineering and regenerative medicine.

With embedded nanomaterials, xenobots could serve as cellular "scaffolds" that guide the growth and repair of tissues.

By acting as bioactive structures, they could stimulate cells to form new tissue in the desired shape, potentially aiding in healing wounds, reconstructing organs, or even regenerating damaged nerves.

Environmental Applications:
Cleaning, Restoring, and Monitoring Ecosystems

Beyond medicine, the merger of xenobots and nanotechnology offers a toolkit for addressing ecological challenges.

Xenobots could be designed to detect and neutralize pollutants in bodies of water, using nanoscale components that target specific contaminants.

For example, xenobots outfitted with magnetic nanoparticles could bind to microplastics or heavy metals, allowing for the targeted removal of these pollutants.

After capturing contaminants, the xenobots could be collected, purified, or even degraded naturally, offering an environmentally friendly solution.

Xenobot-nanotechnology hybrids could also play a role in ecosystem monitoring. These tiny bio-robots, enhanced with nanoscale sensors, could travel through soil, water, or air, collecting data on pollutant levels, nutrient content, or microbial populations.

This data could provide real-time insights into ecosystem health, allowing scientists and policymakers to respond swiftly to environmental threats.

Furthermore, in natural disaster zones, xenobot hybrids could assess environmental damage, like chemical contamination or habitat destruction, enabling informed interventions that could expedite recovery efforts.

Precision Agriculture: Enhancing Crop Health and Yield

In agriculture, xenobot-nanotech hybrids could monitor soil quality, detect plant pathogens, and even deliver nutrients directly to plants. By navigating through the soil, xenobots equipped with nanosensors could assess levels of essential elements like nitrogen and phosphorus.

These systems could then release fertilizers at precisely the right locations and amounts, enhancing efficiency and reducing the environmental impact associated with traditional fertilization techniques.

Additionally, xenobot hybrids could detect pathogens or pests early, allowing farmers to address infestations before they escalate. This precision approach could not only improve crop yields but also reduce the need for pesticides, thus promoting a healthier agricultural ecosystem.

With food security being a critical global challenge, these applications have the potential to transform farming practices, making them more sustainable and responsive to changing environmental conditions.

Future Directions and Ethical Considerations

While the potential of xenobot-nanotechnology hybrids is vast, their development also brings ethical and safety concerns. The release of biohybrid robots into the environment or human body raises questions about unintended interactions with natural organisms or ecosystems.

Moreover, the ability of these systems to self-replicate or self-modify in response to environmental conditions could lead to unpredictable outcomes, making stringent regulation essential.

The fusion of xenobots and nanotechnology also intersects with discussions about privacy, especially in medical or environmental monitoring applications. The use of nanosensors within living organisms could inadvertently collect sensitive data, raising concerns about consent and data protection. Addressing these ethical questions early in the development process will be crucial to ensuring that these technologies are used responsibly.

Bridging the Gap Between Biology and Technology

The merging of xenobots and nanotechnology exemplifies the convergence of biology and technology in ways that were once confined to science fiction.

These biohybrid systems harness the strengths of both fields, creating programmable organisms that could address challenges in medicine, environmental science, and beyond.

Although the development of xenobot-nanotechnology hybrids is still in its early stages, the scientific progress made so far offers a glimpse into a future where biological robots not only perform sophisticated tasks but also do so with a precision that could revolutionize multiple industries.

As we continue to explore the possibilities, the potential for xenobot hybrids to reshape the boundaries of biology and engineering grows.

By responsibly advancing this field, we may soon witness breakthroughs that redefine the role of life and technology in our world.

Creating Xenobots with Microscopic Scales for Specific Tasks

The advent of xenobots, tiny biological machines created from living cells, marks a new frontier in biotechnology, where the boundaries between organic life and synthetic function blur.

These microscopic entities are designed to perform specific, complex tasks at scales smaller than the width of a human hair.

As science seeks ways to optimize xenobots for specialized missions, the challenges of achieving functionality on a microscopic level require innovation across multiple disciplines—biology, engineering, and robotics.

Here, we explore how scientists are creating xenobots at microscopic scales, the technologies enabling these breakthroughs, and the potential applications they hold for medicine, environmental science, and beyond.

Building Microscopic Xenobots: The Science and Tools

Creating xenobots on a microscopic scale involves using biological materials like frog stem cells, carefully harvested and reassembled to function in controlled ways.

Unlike traditional robots with metal and synthetic parts, xenobots are crafted from living tissues, particularly cells from the Xenopus laevis frog.

To engineer them on a microscale, scientists harness micro-manipulation tools and computer-assisted design to guide cells into specific shapes and arrangements that promote desired behaviors, such as movement, sensing, and even rudimentary decision-making.

One of the key aspects of building these microscopic bots is cell selection and specialization. Different types of cells can be programmed to respond in specific ways to stimuli. Muscle cells, for example, can be used to generate motion, while skin cells provide structural integrity.

With the right arrangement and stimulus, these cells can contract, move, and even coordinate with one another at incredibly small scales.

Moreover, recent advancements in synthetic biology have allowed scientists to engineer cells with enhanced properties, such as fluorescence for tracking or magnetism for guiding xenobot swarms.

Microscopic Xenobot Design: Shape, Function, and Scale

On a microscopic level, the design of a xenobot must consider how cells naturally interact with each other, as well as the constraints imposed by physics at small scales. At this level, factors such as surface tension, viscosity, and cellular adhesion can have a huge impact on functionality. For instance, in environments like the human bloodstream, microscopic xenobots face fluid forces that differ significantly from those encountered by larger organisms.

Designing xenobots to overcome these forces requires an understanding of microfluidics—the study of fluid behaviors on a miniature scale.

To address these challenges, scientists are using sophisticated computational models to predict how various xenobot shapes will behave in different environments. By simulating the interactions between cells and their surroundings, researchers can iterate designs before creating physical xenobots.

Computer models can test thousands of configurations and select the optimal shape and cellular composition for specific tasks, such as targeted drug delivery, tissue repair, or micro-scale debris removal.

Some designs include spherical xenobots that are highly maneuverable, while others use elongated forms that can "swim" through fluid by contracting and expanding.

Depending on the task, different shapes are ideal—flattened xenobots might work best for cleaning up cellular debris, while spherical bots can better navigate through fluid environments.

This shape-based specialization is crucial for maximizing efficiency at microscopic scales and enables scientists to tailor xenobot designs to specific challenges.

Applications of Microscopic Xenobots: Targeted and Precise Functions

The development of microscopic xenobots opens up potential applications across a wide range of fields. In medicine, for example, these minuscule bots could be deployed within the human body to deliver targeted therapies directly to diseased cells.

Traditional drug delivery methods are often systemic, affecting large areas of the body, and can lead to unintended side effects. By contrast, xenobots can be designed to navigate directly to the site of an infection or tumor, releasing medication in a localized and controlled manner.

This precision reduces side effects and increases therapeutic efficacy, potentially transforming how conditions like cancer or antibiotic-resistant infections are treated.

Beyond targeted drug delivery, microscopic xenobots hold promise in performing "micro-surgeries" within the body. These tiny constructs could be used to clear clogged arteries, remove small tumors, or even repair cellular damage from within.

Because they are biodegradable and composed of organic cells, they offer a minimally invasive, self-contained method for performing delicate procedures.

In scenarios where traditional surgery is risky or impossible, xenobots present a powerful, less invasive alternative.

Environmental science also stands to benefit from microscopic xenobots. Due to their small size, they could be released into bodies of water to collect pollutants, break down microplastics, or monitor environmental changes at a cellular level.

For instance, xenobots could be deployed in polluted rivers to remove heavy metals or absorb oil droplets, minimizing human intervention and preserving delicate ecosystems.

By tailoring xenobot cells to detect specific pollutants, scientists can create environmental monitoring tools that provide real-time feedback on ecosystem health.

Challenges in Creating and Controlling Microscopic Xenobots

While the potential applications of microscopic xenobots are exciting, achieving consistent control over their behavior and lifespan poses challenges.

Because these bots are biological, their movements and responses can be influenced by unpredictable factors, such as temperature, pH levels, and interactions with other cells.

Additionally, achieving reliable navigation within complex environments like the human body or natural water sources requires a level of precision that current technologies are still working to perfect.

To improve control, scientists are exploring ways to "program" xenobots using genetic engineering, wherein cells can be modified to respond to specific signals. By tweaking cellular pathways, researchers can create xenobots that behave in predictable, repeatable ways.

The use of magnetic fields, electrical signals, and even light as guiding forces is also under investigation, which could allow xenobots to be steered remotely in real time.

However, as xenobots are scaled down to microscopic levels, these external control mechanisms become more difficult to implement effectively. Scientists are thus experimenting with hybrid approaches, combining biological cells with synthetic materials that respond more readily to external cues.

This blending of organic and synthetic components could offer the precise control needed for specialized tasks while retaining the benefits of biodegradability.

Future Prospects and Ethical Considerations

The creation of xenobots at microscopic scales is still in its early stages, and researchers are only beginning to understand the full range of possibilities and limitations. As these capabilities mature, xenobots could redefine our approach to medicine, environmental conservation, and more. However, the ethical implications of deploying microscopic, self-replicating biological machines must also be addressed.

Questions of safety, privacy, and long-term ecological impact are vital, as xenobots could persist in environments or interact with other organisms in unintended ways.

Overall, the pursuit of microscopic xenobots highlights the incredible potential for innovation within biotechnology, as well as the responsibility that comes with creating life forms capable of independent, task-driven behavior. By advancing our understanding of cell behavior at micro scales, xenobot research is pushing the boundaries of what humanity can achieve in both technology and biology—paving the way for new solutions to some of our most pressing challenges.

Chapter 21: Xenobots and Bioenergy Production

In a world seeking sustainable solutions, scientists and engineers are exploring new frontiers, and bioenergy has emerged as a promising area. Among the latest innovations, xenobots — tiny living robots made from biological cells — are being investigated for their potential to contribute to bioenergy production. These unique organisms, developed from frog cells, combine the qualities of engineered design with the natural functionality of living cells.

In the context of bioenergy, xenobots offer exciting possibilities, from biomass harvesting to the transformation of organic matter into energy sources, presenting a blend of biology, technology, and sustainability.

Xenobots are a form of biological robot, initially constructed from the stem cells of Xenopus laevis, an African clawed frog. Unlike traditional robots made of metal or plastic, xenobots are soft and flexible, composed entirely of organic material. They are built from heart and skin cells, with heart cells giving them the ability to move through contractions.

This cellular assembly makes xenobots highly adaptable to various tasks, responding to environmental changes and operating autonomously. Their self-healing and biodegradable nature sets them apart as an innovative tool in fields where synthetic materials are less effective or practical.

Bioenergy and Xenobots: The Potential

The term "bioenergy" refers to energy derived from biological sources, including plant and animal matter. With a focus on sustainability, bioenergy is typically generated through processes like anaerobic digestion, fermentation, or combustion of biomass, which converts organic matter into usable forms of energy, such as biogas or biofuels.

By integrating xenobots into these processes, researchers hope to enhance efficiency and broaden bioenergy's applications.

The potential for xenobots in bioenergy production primarily revolves around their ability to interact with organic material at microscopic levels, enabling them to assist in or speed up energy-producing processes. For example, xenobots could be used to collect and transport organic matter to bio-digesters, where the material is broken down into energy-rich gases. Furthermore, because xenobots are programmable, they could be instructed to seek out specific substances, allowing targeted bioenergy production, enhancing efficiency, and reducing waste.

Mechanisms of Bioenergy Facilitation

One of the critical roles xenobots could play is in aiding microbial ecosystems within bio-digesters. Bio-digesters rely on microbial communities to decompose organic matter, and maintaining the right balance of these microbes is essential for efficient energy production. Xenobots could be programmed to monitor and adjust these microbial communities, ensuring optimal conditions for energy generation.

For instance, they could detect nutrient imbalances and deliver essential compounds to support microbial health and productivity.

In addition to bio-digesters, xenobots may contribute to bioenergy by participating directly in the fermentation process. Fermentation involves the conversion of sugars into gases or alcohols, which are then used as fuels. Xenobots could help by transporting sugar molecules to fermentation sites, optimizing the distribution of materials, and improving overall energy yields. Their precise and small-scale interventions can minimize waste and ensure that every molecule contributes effectively to bioenergy output.

Harvesting Biomass with Xenobots

Another promising application of xenobots in bioenergy is the harvesting of biomass. Biomass, composed of plant material, animal waste, and other organic substances, is an essential raw material for bioenergy production. Gathering this material from natural environments or industrial sites can be challenging and labor-intensive. Xenobots, with their small size and ability to navigate complex environments, could autonomously collect biomass, transporting it to processing sites.

One of the significant advantages of using xenobots for biomass harvesting is their adaptability.

They can move through various terrains and even underwater environments, allowing them to gather biomass in places where traditional methods are less effective.

For instance, xenobots could be released into shallow waters to collect algae, a promising source of biofuel.

Once the algae are gathered, they can be processed to extract oils and other valuable compounds for energy production.

Environmental Monitoring and Waste Reduction

The self-contained, biodegradable nature of xenobots also makes them suitable for environmental monitoring in bioenergy production systems.

As bioenergy often involves managing and processing waste materials, xenobots could be deployed to detect pollutants or harmful byproducts within these systems, ensuring that energy production remains clean and sustainable.

For example, xenobots could monitor pH levels or the presence of toxic substances, signaling when conditions become unfavorable for optimal energy output.

Waste reduction is another area where xenobots could make a difference. In bioenergy facilities, residues and byproducts are often left after energy extraction, and disposing of these materials is a significant challenge.

Xenobots could be used to further break down these residues, either by delivering specific enzymes to accelerate decomposition or by collecting small particles of organic matter for reprocessing.

This approach minimizes waste, maximizing the amount of usable energy derived from raw materials.

Xenobots as Renewable Energy Catalysts

An emerging idea in xenobot research is their potential as catalysts within renewable energy systems. Because xenobots are biological, they can interface naturally with other bioengineered organisms, creating synergies that traditional mechanical robots cannot.

For example, xenobots could be designed to activate specific microbes known to produce methane — a biofuel — when in proximity to organic material. By creating these micro-environments, xenobots could serve as a catalyst for rapid and efficient energy production, operating as "biological reactors" within larger bioenergy systems.

Another catalytic possibility is their use in hydrogen production, a promising energy source. Certain bacteria can generate hydrogen under specific conditions, and xenobots could help by transporting nutrients and managing the micro-environment to optimize hydrogen production.

By assisting these microbial communities, xenobots could contribute to producing clean, renewable hydrogen fuel, which has applications across various industries.

Challenges and Future Prospects

While the potential of xenobots in bioenergy production is remarkable, several challenges remain. One of the primary concerns is controlling xenobots' behavior and ensuring their reliability in complex, real-world environments.

Unlike traditional robots, xenobots operate as living organisms, which means they are subject to biological variability. Programming them to follow specific tasks consistently and accurately will require advances in both bioengineering and artificial intelligence.

Moreover, ethical and regulatory questions about using living organisms for industrial purposes will need to be addressed. As xenobots become more complex and autonomous, regulations will need to ensure that they are safe for the environment and do not disrupt local ecosystems.

Despite these challenges, the future prospects for xenobots in bioenergy production are promising. As research progresses, xenobots could become an integral part of sustainable energy systems, helping to reduce waste, optimize processes, and contribute to a cleaner, greener world.

Their ability to bridge the gap between biology and technology represents a new approach to renewable energy, showing how innovation at the smallest scales can impact the world's energy landscape.

Xenobots and Bioenergy Production

As humanity moves towards greener, more sustainable solutions, the quest for alternative energy sources has gained considerable momentum. One of the more unconventional, yet potentially revolutionary, avenues is bioenergy generated from organic matter.

Imagine a scenario where small, programmable biological machines, known as xenobots, process organic materials and convert them into usable bioenergy. Such a development could redefine how we approach both waste management and energy generation, combining biology and engineering in a way previously limited to science fiction.

Understanding Xenobots and Their Potential for Bioenergy Production

Xenobots are tiny, organic robots created from the cells of African clawed frogs (Xenopus laevis), from which they derive their name. By assembling these living cells in specific arrangements, scientists can program the xenobots to perform simple tasks. Unlike traditional robots, xenobots are soft, biodegradable, and capable of regenerating, making them uniquely suited for tasks within biological environments.

The fundamental structure of a xenobot consists of living heart and skin cells. Heart cells contract and relax, enabling the xenobot to move, while the skin cells provide structure and a certain level of durability.

The simplicity of these cellular structures belies their potential—xenobots have already demonstrated basic forms of locomotion, object manipulation, and even self-replication under specific conditions.

This combination of biological versatility and programmability allows xenobots to interact with their environment in ways that traditional robots cannot. By using xenobots to process organic materials, researchers hope to unlock their ability to break down waste and generate bioenergy, thereby tapping into a sustainable, eco-friendly energy source.

How Xenobots Generate Bioenergy

The idea of using xenobots for bioenergy production centers on their potential to harness chemical reactions from organic matter decomposition. Organic matter, such as plant residues or animal waste, contains complex molecules like carbohydrates, proteins, and fats. When broken down, these molecules release energy in the form of biochemical reactions. Xenobots, being composed of living cells, can be designed to facilitate these reactions, potentially serving as "mini bioreactors."

In this setup, xenobots could interact with organic waste, breaking it down through enzyme-driven processes. Enzymes are biological catalysts that speed up chemical reactions, and cells naturally produce them to process various nutrients.

By programming xenobots to carry enzymes specific to breaking down organic compounds, scientists could effectively design these biological machines to convert waste material into simple, energy-rich molecules like glucose or lipids. This chemical energy can then be harvested and transformed into bioelectricity, biofuel, or other forms of bioenergy.

Bioelectricity from Xenobots

One exciting potential application of xenobots in bioenergy is the generation of bioelectricity, which is essentially electricity produced by living organisms. Bioelectricity is commonly observed in certain bacteria and plants, which produce electrical charges through cellular respiration—a process by which cells generate energy from organic compounds.

By integrating similar bioelectric mechanisms within xenobots, scientists could create mini power cells that convert organic material into electricity.

For instance, xenobots equipped with bioelectrochemical cells could be designed to release electrons during the breakdown of organic compounds.

These electrons would then travel through an external circuit, generating an electric current that could be harnessed as a power source.

This process, if scaled up, might lead to applications where xenobot colonies generate low levels of electricity, potentially powering small devices or contributing to microgrid systems.

Xenobots in Waste-to-Energy Systems

One of the most promising applications of xenobot-based bioenergy production lies in waste-to-energy systems. Currently, waste-to-energy plants use incineration or anaerobic digestion to convert waste into usable energy.

While effective, these methods often have limitations, including the release of greenhouse gases or the need for specific conditions to break down materials.

Xenobots could bring a more environmentally friendly approach to this process. Due to their programmable nature, they can be deployed to target specific types of organic waste, like food scraps or agricultural residues.

The xenobots would then enzymatically decompose the waste into smaller, energy-rich compounds. By collecting and processing these compounds, the xenobots would essentially act as "bio-fuel converters," transforming the waste into biofuels such as methane or ethanol.

Since xenobots are biodegradable and self-sustaining within controlled environments, they could be ideal for closed-loop systems where waste is constantly fed into a xenobot-operated bioreactor.

This approach could drastically reduce the need for external energy inputs and lower emissions associated with conventional waste-to-energy plants.

Environmental and Economic Benefits of Xenobot-Based Bioenergy

If xenobot-based bioenergy systems can be successfully developed, they could bring numerous environmental and economic advantages. Firstly, these systems would offer a significant reduction in waste accumulation.

Organic waste, such as food and agricultural byproducts, constitutes a large portion of municipal and industrial waste, and finding efficient, eco-friendly ways to process it remains a challenge.

Xenobots, by converting this waste into bioenergy, could minimize landfill use and reduce methane emissions—a potent greenhouse gas produced by decomposing organic waste.

Secondly, xenobot-powered bioenergy systems could provide a sustainable, decentralized energy source.

Unlike solar or wind power, which depend on specific environmental conditions, xenobots can function in a variety of settings.

This adaptability makes them particularly useful in remote or underdeveloped regions, where energy resources are scarce but organic material is abundant.

Localized bioenergy systems powered by xenobots could offer a reliable and environmentally responsible energy source for these communities.

Economically, xenobot-based bioenergy could reduce the costs associated with waste disposal and energy production.

Currently, both of these processes require significant infrastructure and energy input.

By integrating xenobots into waste management systems, cities and industries could potentially reduce the operational costs of waste disposal while generating a secondary benefit in the form of bioenergy.

Future Prospects and Challenges

While xenobot-powered bioenergy presents a promising vision, there are challenges to consider. One of the primary obstacles is scalability. Although laboratory studies have shown the potential of xenobots to perform simple tasks, creating large-scale, autonomous bioenergy systems would require extensive development and refinement.

Researchers would need to ensure that xenobots could function in diverse environments, maintain activity over time, and withstand different types of waste materials without degradation.

Furthermore, the ethical and regulatory aspects of deploying xenobots in natural or public settings need thorough examination. As a new form of biotechnology, xenobots bring unprecedented questions regarding safety, ecological impact, and potential misuse.

Ensuring that xenobots can be safely contained and do not disrupt local ecosystems will be essential for gaining public and regulatory support.

Using xenobots to generate bioenergy from organic matter offers an intriguing, innovative approach to sustainable energy production. By leveraging the biological capabilities of these tiny robots, we could open new frontiers in bioenergy that align with environmental goals.

As research advances, xenobot-based bioenergy systems might one day provide a low-cost, eco-friendly, and adaptable energy source, transforming how we manage waste and fuel our world.

Implications for Sustainable Energy Production

Imagine a future where energy isn't extracted from nature but produced, managed, and transported by tiny, self-powered biological machines. This isn't science fiction. Xenobots—microscopic living robots created from frog cells—offer us an exciting glimpse into how biological engineering might revolutionize sustainable energy production. While these tiny, self-replicating organisms are currently experimental, their potential to impact how we generate, store, and distribute energy in a greener, more efficient way is monumental.

Xenobots are a new type of biological robot created by combining stem cells from the African clawed frog (Xenopus laevis). Scientists at Tufts University and the University of Vermont first developed them by using computer simulations to design cellular structures that can be arranged to perform specific tasks. When these designs are applied in the lab, cells self-assemble and begin behaving in ways programmed by their design, essentially turning into tiny biological machines.

Despite their simplicity, xenobots exhibit remarkable behaviors. They can move toward a target, group together, transport small objects, and even self-replicate. These capabilities arise without genetic modification—xenobots are shaped solely through careful arrangement of cells, tapping into biology's existing "toolkit" to create versatile, programmable organisms. Now, scientists are exploring how these tiny robots could address one of our most pressing global challenges: clean, sustainable energy.

Xenobots and Bioenergy

Traditional energy systems rely heavily on finite resources and often produce toxic waste. A bioengineered solution like xenobots, however, offers the promise of renewable energy production with minimal environmental impact. Xenobots run on their own biological "fuel," metabolizing energy from stored lipids and proteins, meaning they're entirely self-sustaining. Imagine a system where these robots could harness solar or chemical energy in natural environments, like oceans or soil, to perform tasks such as breaking down pollutants or converting waste into biofuels.

A promising application of xenobots is the transformation of organic waste into bioenergy. Organic waste, from agricultural residues to food scraps, is abundant yet largely untapped as an energy source. Xenobots could help bridge this gap by acting as "micro-digesters" that break down these materials into usable fuels. Equipped with the right cell types, xenobots could convert waste into biogas or other energy-rich compounds, potentially offering a continuous source of clean energy. Unlike traditional incineration or fermentation processes, xenobots would operate with extreme efficiency at microscopic scales, maximizing energy output while minimizing waste.

Xenobots as Microbial Collaborators

One of the most exciting potentials for xenobots in sustainable energy production lies in their ability to collaborate with other microorganisms. In nature, microbial communities are responsible for numerous energy-producing reactions, from photosynthesis to the decomposition of organic matter. By integrating xenobots into these ecosystems, scientists could amplify or direct these natural processes for energy production.

For example, cyanobacteria—tiny photosynthetic organisms—produce oxygen and other energy-rich molecules by harnessing sunlight. When coupled with xenobots, these microbes could be directed to optimize their photosynthetic processes, producing biofuels or generating hydrogen in much higher yields. Xenobots could "shepherd" these microorganisms, maintaining optimal conditions for energy production by adjusting their environment or removing toxins.

Additionally, xenobots could be designed to help regulate microbial populations, promoting the growth of beneficial species while curbing those that are harmful or unproductive.

Enhancing Solar and Geothermal Energy Systems

Xenobots' biological adaptability makes them ideal for complementing renewable energy systems, particularly solar and geothermal. Solar panels, for instance, often suffer from efficiency losses due to dust and debris buildup. Conventional cleaning methods are water-intensive, costly, and environmentally taxing. Imagine if swarms of xenobots could self-deploy across the surface of solar panels, removing dust and optimizing light absorption. Their autonomous behavior and ability to survive in various environmental conditions would make xenobot-based maintenance systems far more sustainable and economical than traditional methods.

Similarly, geothermal plants could benefit from xenobot technology. By navigating underground water channels, xenobots could be used to unclog geothermal pipes or clear mineral deposits that often hinder energy production. This self-sustaining, self-cleaning approach would reduce downtime and maintenance costs while ensuring more efficient geothermal energy production.

Self-Healing Energy Infrastructures

The regenerative capabilities of xenobots offer a fascinating application in self-healing infrastructures. Energy systems are prone to wear and damage, whether it's a battery degrading over time or a pipeline suffering from corrosion.

With further research, xenobots could become integral to "smart" energy infrastructures capable of repairing themselves.

For example, self-healing solar panels or wind turbine blades could have embedded xenobot colonies that activate when damage occurs, patching small tears or breaks at the microscopic level.

These self-repair mechanisms would extend the lifespan of renewable energy infrastructure, reduce waste, and lower maintenance costs.

For instance, xenobot-infused batteries could have the ability to heal minor defects in the cell structure, thus maintaining capacity and longevity without intervention.

This would be particularly useful in remote installations where regular maintenance is difficult. Imagine off-grid solar storage systems where xenobots manage and repair storage units, keeping them in peak condition even after years of exposure to harsh environments.

Xenobots and Carbon Sequestration

The role of xenobots could also be extended to tackle carbon dioxide (CO_2) emissions directly. Certain microorganisms are naturally adept at capturing and converting CO_2 into biomass, but their processes are often inefficient when scaled up.

Xenobots could potentially be engineered to assist in carbon sequestration, either by shepherding CO_2-consuming microorganisms or by directly engaging in CO_2 absorption. This could be particularly impactful in carbon-intensive industries such as steel and cement manufacturing, where xenobot-assisted systems might capture CO_2 emissions before they're released into the atmosphere.

Additionally, with their small size and flexibility, xenobots could work within various environments, such as soil and ocean sediments, to trap CO_2 and even convert it into organic compounds useful for energy production.

These activities could contribute to a circular carbon economy, where captured CO_2 is fed back into the production of biofuels or other renewable energy sources, closing the loop on carbon emissions.

Challenges and Future Directions

While the promise of xenobots in sustainable energy production is immense, there are considerable hurdles to overcome. Xenobots are a new technology, and their behavior in real-world ecosystems remains largely untested. Their interactions with other organisms, the risk of unintended ecological impacts, and ethical considerations of deploying living machines into the environment must be carefully considered.

Additionally, scaling xenobot production and programming them to perform highly specialized tasks is a scientific challenge that will require years of research and development.

Despite these challenges, the future of xenobots in energy production remains bright. With advancements in bioengineering, machine learning, and renewable energy technologies, xenobots could evolve from laboratory curiosities to integral components of sustainable energy systems.

As researchers continue to explore these living machines' capabilities, the potential for xenobots to create a cleaner, greener world becomes ever more compelling.

Chapter 22: The Dark Side of Xenobots

Imagine a tiny organism, engineered not by nature but by human hands. This creature is neither animal nor machine but a fusion of both, crafted from frog cells and designed to fulfill specific tasks. These are xenobots: the world's first "living robots." First created in 2020 by a team of scientists from the University of Vermont and Tufts University, xenobots captured the world's attention as a new form of bio-engineering that could perform essential, tiny tasks like cleaning microplastics from water or transporting drugs to specific sites in the human body.

But as with all groundbreaking technologies, there's a flip side to the coin. In a world increasingly comfortable with biotechnology, the dark side of xenobots presents ethical, ecological, and existential challenges that could transform our future in ways we aren't yet prepared to understand.

Xenobots are made from the stem cells of the African clawed frog (Xenopus laevis), giving them their name.

Scientists extract skin and heart cells, then use computer models to predict how these cells could be assembled to move autonomously.

Unlike conventional robots made from metals and plastics, xenobots are fully biodegradable, disappearing without leaving a toxic residue. They are capable of self-propulsion, basic tasks, and even self-repair when damaged.

Despite their simple structure, these bots can operate in groups, moving in swarms like cells in the human body.

This ability to function as "bio-machines" is promising for medicine, environmental science, and research. Yet these very qualities also spark fears and raise profound ethical questions.

Potential for Uncontrolled Reproduction and Mutation

One primary concern with xenobots is their potential for uncontrolled reproduction. Although scientists design these bio-robots to be sterile and to have limited lifespans, living cells inherently have the capacity for change and adaptation. This became especially apparent when some versions of xenobots exhibited a surprising ability to self-replicate in a novel way, called "kinetic replication."

In a lab setting, clusters of xenobots nudged loose cells into piles that eventually formed new, self-moving xenobots. While this form of replication differs from that seen in traditional biological organisms, it raises troubling questions about control. If xenobots could, under certain conditions, produce offspring, then they might eventually evolve or mutate in unforeseen ways.

The prospect of engineered organisms reproducing outside a controlled lab setting could lead to an ecological crisis, disrupting ecosystems or competing with natural species in unexpected and harmful ways.

Risk of Ecological Impact

The very qualities that make xenobots attractive for cleaning up environmental pollutants could, in certain scenarios, backfire. Imagine if xenobots intended to biodegrade pollutants were accidentally released or went out of control. They could potentially attack not only pollutants but also natural organic matter essential to ecosystems.

Their presence in oceans or lakes could disrupt entire food chains, endangering marine life and, by extension, human communities that rely on these ecosystems.

Xenobots' ability to swarm might make them highly effective at certain tasks, but this characteristic also makes them difficult to contain once released. Swarming bots could multiply environmental risks in ways scientists have yet to anticipate.

Such disruptions underscore the importance of containing and monitoring xenobots if we're to safely harness their power.

Ethical Implications of Creating New Life Forms

One of the most profound challenges posed by xenobots is ethical. Unlike synthetic chemicals or mechanical robots, xenobots are alive—created from living cells. Their hybrid status raises questions about our moral responsibilities toward them. Should xenobots have rights?

If these organisms possess a basic form of "intelligence" or autonomy, however limited, are they entitled to protection from unnecessary harm?

Furthermore, the act of creating new life forms could set a precedent that accelerates human intervention in natural processes, potentially blurring the line between biological life and artificial constructs.

In pushing these boundaries, humanity might begin to lose sight of what constitutes life, agency, and personhood. The ethical debates surrounding xenobots intersect with philosophical issues around human responsibility, the sanctity of life, and the limits of scientific exploration.

Misuse and Militarization of Xenobots

Another significant fear surrounding xenobots is their potential for misuse. Technologies developed for good can also be weaponized, and xenobots are no exception. Militaries around the world might see the potential of using xenobots as tiny biological spies, capable of covertly entering enemy territory.

Alternatively, xenobots could be engineered to deliver toxins, spread disease, or disrupt ecosystems as a form of biological warfare.

While these scenarios may sound like science fiction, history teaches us that militarization often follows on the heels of innovation. The accessibility and adaptability of xenobots make them particularly vulnerable to repurposing by those who seek to exploit their unique capabilities.

Even if restrictions and regulations are in place, these precautions might not be enough to prevent their use in warfare or espionage.

Privacy Concerns and Surveillance

Another dark side to xenobots involves surveillance and privacy risks. Because they are small, biodegradable, and can move in groups, xenobots could theoretically be used as tools for surveillance.

Imagine swarms of nearly invisible xenobots entering restricted areas or following specific individuals to record their actions without their knowledge.

These bots could even be engineered to carry tiny sensors that transmit information in real time.

While this level of surveillance might aid in monitoring public safety, it also raises concerns about personal privacy and government overreach. As xenobots evolve, they may become increasingly hard to detect, enabling an unprecedented level of covert monitoring that could erode civil liberties.

Unknown Long-term Impacts

Finally, the dark side of xenobots includes a host of unknowns. Because xenobot technology is so new, its long-term effects on human health and the environment remain largely untested.

Scientists can't yet fully predict how these bio-bots will interact with natural cells over time or what unintended mutations might arise in response to various stimuli.

For instance, if xenobots are deployed in large numbers, there's a possibility of cellular cross-contamination that could impact the health of living organisms in the wild or even in human bodies.

These uncertainties underscore the importance of continued, cautious research.

If xenobots develop in ways we can't predict or control, we might find ourselves facing unforeseen consequences that we can no longer reverse.

Moving Forward with Caution

Xenobots are a testament to human ingenuity, pushing the limits of what we thought possible in biotechnology. Yet, as with any innovation, they come with risks that must be carefully managed. Scientists, ethicists, policymakers, and the public all have a role in ensuring that the development of xenobots prioritizes safety, ethical considerations, and environmental stewardship.

Open, transparent discussions about the potential dangers of xenobots are essential, and regulations must be developed to address these risks before they materialize. Ultimately, xenobots invite us to question not only what we can do but also what we should do as we shape the future of life itself.

Addressing Concerns About Potential Misuse of Xenobots

Xenobots, named after the African clawed frog species Xenopus laevis from which they are derived, represent a cutting-edge frontier in biological engineering. These tiny, self-assembling "living machines" are built from frog cells but designed by computers to perform specific tasks. Despite their small size and rudimentary nature, xenobots have stirred up excitement in scientific fields ranging from medicine to environmental science.

However, alongside the fascination with their potential comes a wave of concerns regarding their ethical implications and potential misuse.

Before exploring the potential risks, it's essential to understand what xenobots actually are. Xenobots are neither traditional robots nor conventional organisms. They are bio-hybrid constructs created by assembling stem cells into specific forms using computer algorithms.

The stem cells used typically come from Xenopus laevis embryos, and scientists program the cells into simple clusters that can carry out pre-determined functions, such as movement, object manipulation, or clustering. Unlike traditional robots, xenobots are biodegradable, powered by their own cellular processes, and have a life span of only a few weeks before they break down naturally.

Xenobots represent a blend of synthetic biology and robotics, providing new ways to study life at the intersection of biology, artificial intelligence, and engineering.

Because they're designed with specific tasks in mind, they can potentially tackle environmental challenges, such as cleaning up microplastics, delivering drugs in precise locations within the human body, or even supporting tissue repair.

But just as their applications inspire hope, their unique abilities raise legitimate ethical and safety concerns.

Concern #1: Uncontrolled Replication

One of the major concerns about xenobots is the possibility of uncontrolled replication. Although current xenobots do not reproduce like traditional organisms, recent developments have shown that under certain conditions, xenobots can "replicate" by aggregating loose cells into new xenobot clusters.

This self-replicative behavior is distinct from biological reproduction—it's more akin to a snowball gathering snow as it rolls downhill. While the replication remains constrained within a lab environment, the worry is that future developments could result in xenobots capable of uncontrolled growth in the wild, which could harm ecosystems by competing with natural organisms.

Mitigation Efforts: Scientists working on xenobot development are acutely aware of these concerns. To prevent potential harm, xenobots are designed to degrade naturally within a short period.

Additionally, they're programmed with specific, limited life spans to reduce the risk of any unforeseen proliferation.

Future versions may include additional "safety locks," such as DNA programming that would ensure they can only replicate in specific, controlled environments.

Concern #2: Environmental Impact and Ecological Disturbance

With xenobots' potential as tools for environmental cleanup, there is also the fear that introducing them into natural ecosystems could disrupt local species and balance.

For example, if xenobots were deployed to collect microplastics in the ocean, what would prevent them from inadvertently consuming or affecting microorganisms essential to the ecosystem?

Introducing a synthetic organism into a wild habitat, even with good intentions, might have unpredictable and potentially damaging consequences.

Mitigation Efforts: Environmental deployment of xenobots is still in its conceptual stages, and scientists are aware that any deployment must be approached with extreme caution. Before introducing xenobots to a natural habitat, rigorous trials in controlled environments would assess any potential ecological impact.

Additionally, scientists are designing xenobots that decompose into harmless compounds, ensuring that their breakdown will not introduce foreign or toxic substances into the environment.

Another proposed measure is to use only fully biodegradable materials, which would ensure that xenobots leave no lasting traces once their tasks are complete.

Concern #3: Biosecurity Risks

Xenobots could also present biosecurity risks if used maliciously. Hypothetically, xenobots could be engineered for harmful purposes—perhaps to carry toxins or disrupt biological processes in other organisms.

This "dual-use" potential is common in advanced technologies but becomes especially concerning with xenobots, as they're both programmable and biological.

Mitigation Efforts: To address biosecurity risks, scientists and regulators are establishing ethical guidelines and legal frameworks to prevent malicious use. Xenobot research typically occurs in secure laboratories under strict oversight, and any work involving xenobot modifications for medical or environmental applications requires approval by regulatory agencies.

Further, many researchers advocate for open, transparent communication about xenobot research, which could enable the scientific community to more readily spot and counter any misuse.

Concern #4: Ethical Considerations Surrounding "Living Machines"

Xenobots challenge our notions of life and autonomy. While xenobots lack nerves, brains, or any capacity for consciousness, some people question whether creating living machines from cellular material is ethically appropriate. Do we risk venturing into morally gray areas by manipulating life in this way? Moreover, as xenobots evolve, might we reach a point where they possess more complex biological functions that blur the line between a machine and an organism?

Mitigation Efforts: Ethical oversight in xenobot research is a top priority. Institutional review boards and ethics committees evaluate research proposals to ensure that xenobot experiments comply with ethical standards. Additionally, transparency with the public about the purpose and limitations of xenobots can help alleviate concerns. Open discussion on ethical boundaries in synthetic biology is an essential part of research, aiming to build societal trust and establish consensus on responsible usage.

Concern #5: Privacy and Surveillance Risks

Though far from achieving human-like intelligence or surveillance capabilities, xenobots could theoretically be modified to collect data on their surroundings, such as environmental conditions or biological markers. This could have implications for privacy, especially if xenobot technology advances to a point where they could collect data on human health or behavior.

Mitigation Efforts: Current research does not involve xenobots that gather private data, and stringent regulations would apply to any such applications. If future xenobots were to be used in data-gathering roles, it would likely be in controlled medical settings, where strict privacy safeguards would be in place. Additionally, public policy discussions and frameworks could help establish clear boundaries on data collection capabilities.

Looking to the Future

The development of xenobots presents both incredible opportunities and challenges. Addressing the concerns surrounding xenobot misuse requires a balanced approach that combines strict regulatory oversight, open scientific dialogue, and responsible engineering practices. By doing so, society can maximize the benefits of xenobot technology—such as targeted drug delivery, environmental remediation, and biocompatible machines—while minimizing risks.

As with any emerging technology, xenobots offer a mirror to our values and ethical priorities. Their responsible development will depend on our willingness to question, challenge, and innovate with caution. The future of xenobots promises a glimpse into an era where biological machines work in harmony with nature—provided that the scientific community and society at large continue to work collaboratively in addressing concerns about potential misuse.

Safeguarding Against Unintended Consequences

As we embark on a journey into the frontier of synthetic biology, the development of xenobots — tiny, programmable organisms made from living cells — promises both extraordinary possibilities and critical questions. Xenobots are living machines, engineered by combining the principles of biological science and artificial intelligence, enabling them to move, perform tasks, and even self-repair. While their applications could revolutionize medicine, environmental science, and robotics, they also raise vital concerns about unintended consequences. Ensuring the responsible development and deployment of xenobots requires a proactive approach to risk assessment, regulation, and ethical stewardship.

Understanding Unintended Consequences in Xenobot Technology

In any transformative technology, unintended consequences are often a byproduct of the complexities and unpredictabilities inherent in novel systems. In the case of xenobots, unintended consequences could manifest in various ways, from ecosystem disruption to unanticipated interactions with other biological systems.

Since xenobots are living organisms, they have unique capacities — such as growth, adaptation, and mutation — that may produce effects beyond human control. While their small scale and limited capabilities suggest they would have a minimal impact, we must consider potential risks that emerge as they evolve or are scaled up in applications.

1. Ecological Impact and Containment

Xenobots are designed to interact with their environment, potentially for applications like cleaning up microplastics, delivering targeted medicine, or repairing tissues. However, when living organisms are introduced into new environments, the ecological consequences can be complex and difficult to predict.

One potential concern is that xenobots, in their pursuit of performing intended tasks, could interfere with natural ecosystems. They might unintentionally disrupt microbial communities, harm beneficial organisms, or introduce foreign genetic material into the ecosystem.

Containment strategies are essential to prevent xenobots from escaping into unintended environments. These could include genetic "kill switches" that deactivate xenobots if they stray beyond specific conditions or environments.

Additionally, engineering xenobots to require synthetic nutrients or environmental triggers that are absent in nature could make them less likely to survive or replicate outside controlled settings.

By implementing these safeguards, researchers can minimize the risk of xenobots inadvertently affecting ecosystems.

2. Genetic Stability and Mutation Control

A xenobot's potential for mutation, although low in their current forms, remains an area of concern as the technology evolves. Mutation is a natural process in all living organisms and can lead to unintended traits or functions in xenobots over time.

This could mean the gradual drift from the xenobot's original design, altering its behavior or lifespan in unpredictable ways.

To safeguard against such genetic instability, researchers can use specific cellular types that have a lower likelihood of mutation or employ gene-editing techniques that make the xenobot's genome more resistant to change. Alternatively, researchers can limit xenobots' lifespans, ensuring they degrade before any substantial mutations occur.

By emphasizing genetic stability, scientists can help prevent xenobots from deviating from their intended design, preserving control over their functions and behavior.

3. Human Health and Safety

The interaction between xenobots and human biology, particularly in medical applications, presents both promising opportunities and potential risks. Xenobots might one day be used to deliver targeted therapies, clear harmful bacteria, or even help in regenerating damaged tissues.

But these applications require a precise understanding of how xenobots interact with human cells and immune systems. Unexpected immune reactions, inflammatory responses, or adverse interactions could pose health risks if xenobots were to behave unpredictably.

Testing xenobots rigorously in controlled settings is essential before any medical application. Researchers could design xenobots that biodegrade or deactivate after completing their tasks, reducing the risk of lingering effects on the body. Furthermore, advanced monitoring and regulatory frameworks can ensure that xenobot deployment in humans undergoes extensive scrutiny, minimizing health risks and ensuring that these innovations truly benefit patient care.

4. Data Privacy and Ethical Considerations

As xenobots may one day be able to collect biological data in the human body or the environment, this raises concerns about data privacy and ethical considerations. If xenobots are programmed to sense, analyze, or transmit data, there could be risks regarding who has access to the information, how it is stored, and what it is used for.

In medical applications, for example, xenobots might collect sensitive health data, raising questions about patient consent and the protection of personal information.

To mitigate these issues, it's essential to design robust data governance protocols and privacy measures tailored to xenobot applications. Legal and ethical frameworks should outline the acceptable uses of data gathered by xenobots, emphasizing transparency, patient consent, and data security.

Additionally, restricting xenobots from transmitting data to external devices or networks could further safeguard privacy.

5. Long-Term Evolution and Self-Replication Risks

Xenobots currently lack the ability to self-replicate; they are intentionally designed as limited-lifespan organisms with specific tasks. However, future advancements in bioengineering may give rise to self-replicating or self-sustaining xenobots, raising concerns about control and long-term evolution. Self-replicating xenobots, if ever developed, could evolve, leading to unpredictable changes in behavior or function.

In a worst-case scenario, they might become invasive in certain environments, outcompeting native organisms or impacting biodiversity.

Strict regulatory oversight and containment protocols would be critical if self-replicating xenobots were ever considered.

Researchers could design mechanisms that limit replication to a finite number of cycles, ensuring that xenobots degrade or deactivate after a set period.

Additionally, including genetic "safeguards" or control sequences that prevent uncontrolled replication could further reduce these risks. As we approach the potential for self-replication, it will be essential to weigh the benefits against the risks with an ethical, cautious approach.

6. Responsible Innovation and Regulation

Safeguarding against unintended consequences requires more than technical solutions; it demands a commitment to responsible innovation and regulation. Xenobot development should be governed by international standards, informed by experts in synthetic biology, ethics, ecology, and public health. Regulatory bodies could establish protocols for risk assessment, transparency, and accountability, ensuring that the benefits of xenobots do not come at the cost of environmental safety or public health.

Incorporating public opinion and fostering open communication about xenobot technology can also play a significant role in establishing trust. Transparent reporting of xenobot research, potential risks, and ethical considerations can help engage the public in meaningful dialogue, addressing concerns and establishing a shared sense of responsibility.

The Path Forward

As we stand at the threshold of a new era with xenobots, the stakes are high. These tiny living machines offer immense potential, but without careful consideration, they also pose real risks. By focusing on containment, genetic stability, human safety, data privacy, and ethical stewardship, we can strive to navigate the uncharted waters of xenobot technology with caution.

Safeguarding against unintended consequences is not just about preventing negative outcomes; it's about shaping a responsible future where xenobots serve humanity without compromising our ecosystems or ethical principles.

Embracing a thoughtful, anticipatory approach to xenobot development will help ensure that these innovations remain a positive force, advancing science while respecting life and preserving the natural world.

Chapter 23: Xenobots in Agriculture and Environmental Cleanup

The development of xenobots—tiny, programmable living organisms created from frog cells—has opened doors to fascinating possibilities in diverse fields. These remarkable bio-bots were first developed by researchers at Tufts University and the University of Vermont in 2020, crafted using stem cells from the African clawed frog (Xenopus laevis), from which they get their name.

These unique bio-machines can move, work together, and perform simple tasks without complex brain structures, using only the natural capabilities of their cells.

Beyond the initial marvel of their creation, scientists are exploring ways these biological robots could benefit humanity by enhancing environmental sustainability and revolutionizing agriculture.

Xenobots in Agriculture:
Precision and Environmental Responsibility

Xenobots have the potential to become invaluable assets in modern agriculture.

By deploying tiny, organic machines within agricultural systems, scientists envision methods to promote crop health, enhance soil fertility, reduce pesticide use, and increase crop yields—all while minimizing environmental impacts.

1. Soil Health and Crop Monitoring

The health of soil is crucial for sustainable agriculture, as healthy soil is the foundation of productive, resilient ecosystems.

Xenobots could be introduced into agricultural fields to monitor soil quality in real-time, providing farmers with essential information on moisture levels, nutrient distribution, pH balance, and the presence of toxic substances.

Traditional soil monitoring methods often require invasive procedures and expensive equipment. In contrast, xenobots offer a non-invasive, biodegradable solution. These tiny robots could continuously monitor soil conditions and move autonomously to regions where attention is needed.

By monitoring soil health, xenobots could prevent issues before they arise. If soil pH levels begin to shift, xenobots could trigger an alert, allowing farmers to address the problem quickly before it impacts crop health. By gathering and analyzing data from the field, xenobots could also aid in targeted fertilization, ensuring that nutrients are applied only where necessary, reducing chemical runoff into water systems and lowering overall fertilizer use.

2. Pest Control with Minimal Pesticides

In conventional agriculture, the use of pesticides and insecticides is widespread, yet these chemicals often harm non-target species, pollute waterways, and contribute to the decline of beneficial insects like bees and butterflies. Xenobots could serve as a biological form of pest control, offering an eco-friendly alternative to traditional pesticides. Scientists are investigating ways to program xenobots to identify and target specific pests, potentially allowing them to interact with pest populations while avoiding other, beneficial organisms.

This precision-based approach could reduce the need for broad-spectrum insecticides and other chemicals, preserving biodiversity within ecosystems and preventing harmful pesticide residues from entering the food chain.

By employing xenobots in pest management, farmers may be able to reduce crop damage without the adverse environmental effects associated with chemical pest control.

3. Biodegradable Field Operations

Xenobots are naturally biodegradable, breaking down harmlessly after completing their tasks. This quality makes them ideally suited for agricultural tasks that require minimal long-term environmental impact. Unlike plastic-based agricultural drones or heavy machinery, xenobots do not leave behind non-biodegradable residues that contribute to soil and water pollution.

Their ability to decompose harmlessly allows them to perform tasks that would otherwise require costly cleanup efforts, enhancing the sustainability of farming practices.

Over time, as these xenobots fulfill their intended purposes, they will naturally decompose, contributing to the organic matter in the soil. This process helps maintain soil health while reducing the environmental footprint of agricultural interventions.

Xenobots in Environmental Cleanup: Addressing Pollution

Beyond agriculture, xenobots show great promise in environmental cleanup, specifically in addressing pollution in both terrestrial and aquatic environments. Pollution is one of the most pressing environmental issues today, impacting ecosystems and human health.

Traditional cleanup methods are often labor-intensive, costly, and sometimes ineffective. Xenobots offer a potentially transformative solution, combining the adaptability of biology with the precision of programming to target and neutralize pollutants.

1. Microplastic Removal from Water Bodies

Microplastic pollution has become a pervasive environmental issue, affecting oceans, rivers, and lakes worldwide. These tiny plastic particles are challenging to remove using conventional filtration techniques due to their size and dispersion. Xenobots could provide an innovative solution. Researchers are investigating ways to program xenobots to detect and capture microplastics, even in large water bodies.

Because xenobots are small and self-motile, they could navigate water environments to seek out microplastics and gather them for easy removal. As biodegradable organisms, they do not pose a threat to aquatic life. After collecting microplastics, they could be collected and safely decomposed, with both the xenobots and pollutants removed from the ecosystem.

2. Oil Spill Cleanup and Pollution Control

Oil spills are devastating to marine ecosystems, contaminating water, harming wildlife, and impacting coastal communities. Current cleanup methods, like dispersants and oil skimmers, can only achieve partial success and often lead to secondary environmental problems.

Xenobots could potentially revolutionize oil spill cleanup efforts. By programming xenobots to detect and bind with oil particles, scientists aim to create a system that can precisely target and collect oil without dispersing it further.

As xenobots swarm over an oil spill, they could gather oil droplets, which could then be collected for safe disposal or reprocessing. The use of xenobots for oil cleanup has the potential to be more effective than current technologies and would leave minimal ecological impact, as they biodegrade after completing their mission. This approach could be particularly beneficial for smaller spills that are challenging to address with conventional means.

3. Decomposition of Chemical Pollutants

Industrial activities release numerous chemicals into the environment, many of which are toxic and persistent. These pollutants accumulate in soil and water, disrupting ecosystems and posing health risks to humans and animals alike. Xenobots offer a targeted method for addressing such pollutants.

Researchers are exploring the use of xenobots programmed to recognize and decompose specific chemicals, such as heavy metals or persistent organic pollutants, neutralizing them before they spread further into the environment.

For example, xenobots could be engineered to detect and bind to molecules of a pollutant like mercury or lead, breaking it down into less harmful components.

This would allow contaminated areas to be detoxified without the need for large-scale, invasive cleanup methods. By effectively containing and breaking down chemical pollutants, xenobots could play a significant role in restoring and preserving natural ecosystems.

Challenges and Ethical Considerations

While the potential of xenobots in agriculture and environmental cleanup is vast, their deployment comes with challenges and ethical questions. Concerns include ensuring xenobots remain contained to their target areas, preventing unintended ecological impacts, and establishing regulations that govern their use.

Researchers are committed to studying the long-term ecological implications of xenobot technology, aiming to strike a balance between innovation and environmental stewardship.

Xenobots represent an exciting step forward in sustainable technology. Their unique combination of programmability, biodegradability, and biological function offers new ways to address pressing environmental and agricultural challenges.

As research progresses, these living machines may become a vital tool for improving soil health, managing pests, removing pollutants, and promoting environmental sustainability on a global scale.

How Xenobots Are Set to Revolutionize Agriculture and Soil Health

In the world of agriculture and soil health, the latest advancements in biotechnology are sparking significant intrigue.

Among them are xenobots, a groundbreaking new development that could redefine how we approach farming and sustainable soil management.

Xenobots are neither traditional robots nor fully organic organisms; rather, they are an entirely novel type of lifeform created through the assembly of living cells, often using frog embryos. With a structure and function designed for specific tasks, xenobots can autonomously carry out activities that conventional technology and biological organisms struggle to perform effectively. This promising technology holds the potential to address some of the most pressing challenges in agriculture, including enhancing soil health, promoting sustainable farming, and reducing the environmental impact of traditional agricultural practices.

Xenobots, named after the African clawed frog Xenopus laevis from which their cells are derived, are programmable, living "robots" made from organic materials. Developed by researchers at Tufts University and the University of Vermont, xenobots are typically created by sculpting embryonic stem cells into shapes that can perform basic functions like movement, healing, and carrying tiny payloads. Unlike mechanical robots, xenobots are biodegradable and rely on natural cell interactions to move and respond to environmental cues. This unique property allows them to interact with their surroundings in a way that is safe, sustainable, and adaptable.

These characteristics make xenobots particularly suitable for agricultural applications. Since they are made of organic cells, they pose minimal risk of pollution and have a limited lifespan, ensuring that they break down naturally over time.

Additionally, xenobots can be engineered to perform specific functions such as detecting chemical changes in soil, decomposing organic matter, or even transporting beneficial microbes to plant roots. This adaptability opens up exciting possibilities for their use in maintaining soil health and improving agricultural efficiency.

How Xenobots Enhance Soil Health

Healthy soil is the foundation of successful agriculture, but modern farming practices can deplete essential nutrients, increase erosion, and disrupt the balance of beneficial microorganisms. Here, xenobots offer a novel solution by working directly within the soil to monitor, maintain, and rejuvenate its composition.

Below are some key ways in which xenobots could contribute to healthier soil:

1. Soil Microbiome Management: The soil microbiome, a complex ecosystem of microorganisms, plays a critical role in plant growth by assisting in nutrient absorption and protecting crops from pathogens.

Xenobots could be deployed to transport beneficial microbes directly to plant roots, fostering a more favorable environment for plant growth.

By introducing these helpful organisms to areas where they are most needed, xenobots could help counteract the effects of soil degradation and reduce the reliance on chemical fertilizers.

2. Decomposition and Nutrient Recycling: Organic matter, such as plant residues and animal waste, must be decomposed to release essential nutrients back into the soil. Xenobots can assist in this process by breaking down these materials at a microscopic level, accelerating nutrient recycling.

Their small size allows them to access areas that larger decomposers, such as earthworms, cannot reach, enabling more efficient nutrient distribution.

By enhancing decomposition processes, xenobots can improve soil structure and fertility, ultimately boosting crop yields and reducing the need for synthetic fertilizers.

3. Pollutant Detection and Bioremediation: Soil pollution from pesticides, heavy metals, and other contaminants is a growing concern for farmers.

Xenobots could be engineered to detect and neutralize harmful substances in the soil, effectively "cleaning" the soil environment. By identifying pollutants on a molecular level, xenobots could alert farmers to areas of concern and potentially detoxify small areas through natural processes. This ability could reduce the accumulation of harmful chemicals in the food chain and help restore contaminated land to productive use.

4. Erosion Prevention: Soil erosion is a major challenge in agriculture, leading to nutrient loss and degraded land. Xenobots might be used to support soil stability by interacting with soil particles and binding them together, helping to reduce erosion. Through targeted movement and distribution of natural binding agents, xenobots could improve soil cohesiveness, particularly in areas prone to erosion due to wind or water runoff. This approach could help conserve soil resources, especially on sloped or arid lands.

Improving Water Management and Efficiency in Agriculture

Efficient water use is critical in agriculture, especially in regions facing water scarcity. Traditional irrigation methods can be wasteful, and overwatering can lead to issues like soil compaction and nutrient leaching. Xenobots could help optimize water management by monitoring moisture levels in the soil and alerting farmers when irrigation is needed. They could also be used to guide water to specific plant roots, reducing the overall water consumption of crops. This approach would enable more precise water distribution, potentially conserving water resources and reducing the environmental impact of agriculture.

Xenobots could also interact with soil particles to improve water retention in dry soils. By promoting better soil structure, they could enhance the soil's ability to retain moisture, making crops more resilient to drought conditions. This application could prove invaluable in regions where water scarcity is a critical concern, allowing for sustainable farming practices even in challenging climates.

Supporting Sustainable Farming Practices

In the context of sustainable agriculture, xenobots offer an alternative to chemical-intensive farming methods. By providing natural means of soil enrichment and pest control, xenobots could help reduce dependence on chemical fertilizers, pesticides, and herbicides. This reduction in chemical usage would not only lower farming costs but also minimize the ecological impact of agriculture, helping to protect biodiversity and reduce soil and water contamination.

Xenobots could be tailored to target specific agricultural pests, acting as biological control agents that do not harm beneficial insects or plants. This precision would enable farmers to manage pests more effectively while avoiding the unintended consequences of broad-spectrum pesticides. In turn, healthier crops and fewer chemical residues could lead to improved food safety and reduced environmental degradation.

The Future of Xenobots in Agriculture

While xenobot technology is still in its infancy, ongoing research suggests that these living robots have the potential to transform agriculture in a meaningful way.

However, several challenges remain before they can be widely adopted. Ethical considerations, regulatory oversight, and scalability are key areas that need to be addressed as scientists work to refine xenobot capabilities and explore their applications in real-world environments.

Additionally, the cost of developing and deploying xenobots needs to be reduced to make them accessible to farmers of all scales, including smallholders in developing regions.

Despite these challenges, the potential benefits of xenobots in agriculture and soil health are too significant to ignore. By merging biology and robotics, xenobots offer an innovative, environmentally friendly approach to farming that aligns with the principles of sustainability and resource conservation. If successfully integrated into agricultural systems, xenobots could help secure a more resilient and productive future for global food production.

In the next few decades, xenobots may become indispensable allies in the quest for sustainable agriculture, offering farmers a powerful tool to manage soil health, enhance crop yields, and protect natural ecosystems. As this technology continues to evolve, xenobots could pave the way toward a more efficient, sustainable, and environmentally responsible agricultural industry.

The possibilities are as exciting as they are transformative, marking a new chapter in the chronicles of agricultural science.

Cleaning Up Pollutants and Contaminants Using Xenobots

Imagine a tiny, programmable machine made from living cells, capable of moving through water or soil, finding pollutants, and breaking them down naturally. Xenobots, a breakthrough in bioengineering, are the first organisms of their kind—part robot, part biological life.

Derived from the skin and heart cells of frogs (the African clawed frog Xenopus laevis, hence the name "xenobot"), these living machines are designed to clean up pollutants and contaminants in a way that is sustainable, safe, and eco-friendly. But how do these strange creations work, and how can they help clean our planet?

Xenobots are millimeter-sized "biobots" crafted by scientists from biological tissue. They are different from traditional robots, which are usually made of metal, plastic, or other inorganic materials. Instead, xenobots are composed entirely of living cells, taken from frog embryos and then carefully manipulated into specific shapes using computer-designed algorithms.

Once shaped, these bundles of cells display programmed behaviors, like swimming through water, moving along specific paths, or even grouping together to carry tiny objects.

These capabilities make xenobots ideal for tasks in environments that traditional robots or chemicals cannot navigate as efficiently or safely, such as polluted rivers, oceans, or even complex soil structures. Since they are biodegradable and non-toxic, xenobots naturally decompose after completing their tasks, leaving no additional waste behind.

Xenobots as Pollution Cleaners

One of the most promising applications of xenobots lies in their potential to clean pollutants and contaminants from water, soil, and other environments. Pollutants like plastics, heavy metals, and chemical waste can pose long-term hazards to wildlife and ecosystems. Xenobots could be programmed to navigate and target specific pollutants, breaking them down or transporting them to designated areas for collection and disposal.

How Xenobots Can Target Pollutants:

1. Swimming and Navigating: Xenobots can move independently, using the natural contractile force of their heart cells to propel themselves forward. With specific shapes and forms, xenobots can travel through water systems, soil, or even confined spaces, where they might encounter small-scale pollutants that other methods cannot access.

2. Sensing Pollutants: By engineering xenobots with certain receptors or chemical signals, scientists can design them to be attracted to or repelled by certain pollutants.

For example, xenobots could be directed to areas with high concentrations of oil or plastic particles, helping them effectively "seek out" and target these contaminants.

3. Absorbing and Breaking Down Contaminants: Certain xenobots could potentially be engineered to contain enzymes or other cellular components that break down specific pollutants.

For instance, they could be designed to carry enzymes capable of decomposing plastic molecules or oil compounds into less harmful substances. In other cases, xenobots may transport these pollutants to a collection point, where they can be safely processed.

4. Collecting Microplastics: Microplastics are tiny pieces of plastic that are difficult to filter from water systems but are harmful to aquatic life and, ultimately, human health.

By leveraging the xenobots' ability to "herd" or move in specific patterns, scientists can design xenobots that capture or corral microplastics for collection. Over time, this could significantly reduce the number of microplastics contaminating oceans, lakes, and rivers.

Advantages of Using Xenobots for Pollution Control

Xenobots offer several advantages over conventional methods of pollution control, especially when dealing with delicate or remote environments.

1. **Biodegradability:** Unlike conventional machinery or chemical treatments, xenobots are entirely biodegradable. After completing their task, they naturally break down, leaving no toxic residue or waste, which makes them safer for ecosystems. This is especially valuable in sensitive areas like coral reefs or freshwater systems where traditional cleanup efforts might inadvertently cause harm.

2. **Energy Efficiency:** Traditional robots or cleanup machines often require external power sources or batteries, which can be costly and limit their mobility. Xenobots, on the other hand, are powered by the biological energy stored in their cells. They can sustain themselves for weeks without needing additional fuel, making them highly efficient.

3. **Precision:** Xenobots can be precisely programmed to operate in specific areas or target specific pollutants. This level of customization means that xenobots can perform targeted cleanups in locations that might be hard to reach otherwise, such as underwater crevices, small rivers, or polluted estuaries. Their small size also allows them to operate without disturbing the surrounding environment.

4. **Self-Replication and Adaptability:** Although still under study, certain xenobot designs may potentially exhibit limited forms of self-replication, meaning they could produce "offspring" xenobots from available biological material in their environment.

If this process can be controlled, xenobots might one day adapt to ongoing pollution issues, continuously cleaning affected areas with minimal human intervention.

Challenges and Ethical Considerations

While xenobots represent an exciting leap forward in pollution management, several challenges remain. For one, xenobots currently have a limited lifespan and need to be replaced periodically.

Additionally, while they can perform specific tasks, their behaviors need to be carefully controlled to avoid unintended impacts on ecosystems. There are also concerns about introducing engineered organisms into the wild, even if they are designed to be biodegradable and non-replicating.

Ethical questions also arise concerning the creation and deployment of xenobots. Since they are derived from living cells and programmed to perform specific tasks, some people wonder where xenobots fit within the definitions of life and machinery. Are they a new form of life, or simply programmable tools? As xenobots continue to develop, ethical guidelines will need to ensure that they are used responsibly, without causing harm to natural ecosystems or compromising environmental integrity.

The Future of Pollution Cleanup with Xenobots

The potential for xenobots in pollution cleanup is immense. Researchers envision a future where fleets of xenobots could be deployed in polluted waters or across farmlands to break down agricultural chemicals, clean oil spills, and even remove heavy metals from soils. These xenobot "swarms" could communicate and work in unison, forming a kind of intelligent, self-sustaining cleanup crew for our planet.

In the coming years, scientists hope to enhance xenobot capabilities by improving their lifespan, refining their ability to target specific pollutants, and possibly even adding new functions. For example, xenobots might one day be capable of detecting and neutralizing hazardous chemicals or pathogens, reducing the risk of contamination in drinking water sources.

A Greener Planet with Xenobots

Xenobots represent a unique fusion of biology and technology, holding the promise of a cleaner, healthier planet. As we face increasing challenges related to pollution and environmental degradation, xenobots offer a sustainable alternative that works harmoniously with nature rather than against it.

These tiny, biodegradable machines are more than just scientific curiosities—they're pioneers in a new era of eco-friendly technology, demonstrating how biological innovation can help restore the balance in our ecosystems. With ongoing research, xenobots could play a pivotal role in the future of pollution control, ultimately helping us protect the Earth for generations to come.

Chapter 24: Public Perception and Xenobot Acceptance

Xenobots, living robots crafted from frog cells, have captivated the world since their introduction. These tiny, programmable biological machines promise revolutionary advances in medicine, environmental science, and technology. Yet, their potential raises questions about ethics, safety, and the ways humans might integrate them into society. As we navigate the early stages of this field, public perception and the social acceptance of xenobots are central to their development and deployment.

The Roots of Skepticism

The concept of creating "living robots" can evoke strong emotional reactions. When people hear about self-moving, self-healing biological entities, they often experience a mixture of awe, curiosity, and anxiety. Some fear that xenobots might lead to uncontrollable biological creations, while others worry about ethical implications of manipulating life. These concerns are deeply rooted in the idea that life forms—even those designed with clear purposes—may behave unpredictably. This fear is compounded by decades of science fiction and popular media that portray artificial life or biotechnology gone awry, further fostering an inherent skepticism.

Moreover, the fact that xenobots are created from animal cells (specifically, the African clawed frog, Xenopus laevis) adds another layer of concern. The thought of crafting organisms from animal tissue touches on issues of animal welfare, as well as philosophical questions about the nature of life itself. People may ask, "Is this ethical?" or "Could these tiny biological machines somehow suffer?" Researchers have addressed these issues by explaining that xenobots are simple, undifferentiated cells without a nervous system, making them incapable of feeling pain or experiencing conscious thought.

However, communicating this distinction remains a challenge as the public grapples with the ethical dimensions of xenobot science.

Ethical Concerns and Transparency

One major factor influencing xenobot acceptance is the ethical debate surrounding their use and development. Bioethics experts, policymakers, and scientists are exploring questions about how far we should go in creating life forms, even if they are entirely designed for beneficial purposes.

To many people, xenobots represent a step into a morally gray area. The act of creating an organism purely for human use and manipulation can be unsettling to those who view it as "playing God" or crossing an invisible line in science.

Addressing these concerns through transparency and public dialogue is essential. Scientists are making efforts to communicate their intentions clearly, explaining that xenobots are designed not for harm but for a range of helpful applications, such as delivering targeted medicine, cleaning up environmental pollution, and repairing tissues.

Through such transparency, they hope to foster a sense of trust and demonstrate that xenobot research is driven by a commitment to human and environmental welfare. The more the public understands the science and motivations behind xenobots, the more comfortable they may become with their development.

Media's Role in Shaping Perception

The media plays a powerful role in how xenobots are perceived. Headlines with terms like "living robots," "self-healing machines," and "biological AI" can trigger excitement but also misunderstanding and fear.

Media coverage has sometimes emphasized xenobots' potential for autonomy or even presented them as "alive" in ways that conflate them with more complex organisms.

When xenobots are described as "programmable life" or compared to science fiction's robotic entities, this can create an exaggerated image of their capabilities.

As a result, some people may mistakenly believe xenobots have abilities or risks that they do not. Xenobots, in reality, are small clusters of frog cells programmed to perform simple tasks within controlled environments. They do not have brains, cannot replicate in a typical biological sense, and lack the capacity for complex behavior.

Educating the public on these distinctions—through careful media communication—will be essential for building an accurate understanding and reducing misconceptions that could hinder acceptance.

The Promise of Xenobots in Health and Environment

As awareness about the tangible benefits of xenobots grows, public opinion may become more favorable. In medicine, xenobots hold the promise of groundbreaking innovations. They could one day be used to deliver drugs precisely to affected areas, reducing side effects associated with conventional treatments.

Furthermore, xenobots might help surgeons repair tissues by navigating tiny spaces within the human body, promoting more effective healing and minimizing invasive procedures.

Such applications have the potential to save lives and reduce suffering, which could significantly enhance public acceptance.

Xenobots also offer exciting potential in environmental science. Imagine armies of tiny, biodegradable robots that can collect microplastics from water sources or clean up toxic waste spills.

Since xenobots are made of organic cells, they biodegrade naturally, posing little long-term environmental threat.

This aligns with the global shift toward sustainable technologies, a factor that resonates positively with many people.

As the public witnesses tangible improvements to health and environmental well-being, resistance to xenobots may diminish, with people recognizing them as tools for the greater good.

Education and Social Engagement

Acceptance of xenobots will likely depend on effective education and engagement with the public. Workshops, interactive presentations, and open discussions between scientists and citizens could serve as powerful tools for alleviating fears and dispelling myths.

Additionally, incorporating xenobot studies into school and university curricula can give young people a balanced understanding of their benefits and limitations, fostering a new generation that is both informed and cautious.

In the future, scientists and institutions may even host public "meet-a-xenobot" events, allowing people to see these tiny robots up close, learn about how they work, and ask questions directly. Such experiences could humanize xenobot technology, shifting perception from an abstract concept to a practical, comprehensible tool. Public engagement fosters an environment of mutual respect and understanding, which could lead to a more informed and accepting populace.

Regulatory and Safety Considerations

Clear regulatory frameworks are crucial for fostering public trust. People are more likely to accept xenobot technology if they believe it is being developed within strict ethical and safety guidelines. Governments and scientific organizations are already working to set these standards, ensuring that xenobots are not used irresponsibly or released into uncontrolled environments.

These regulations may require xenobot researchers to conduct rigorous testing before deploying them, especially if their applications involve direct interaction with humans or ecosystems.

By establishing well-structured guidelines, regulatory bodies can help reassure the public that xenobot technology is safe and beneficial. Moreover, transparent communication about these regulations, paired with accessible information about xenobot studies, could reinforce the public's confidence that this new technology is being handled responsibly.

Toward a Future of Acceptance

While xenobot technology is still in its infancy, its potential to transform medicine, environmental care, and other fields is immense. If we continue to communicate openly, involve the public in discussions, and prioritize ethical considerations, xenobots may be accepted as valuable tools for society.

This journey, however, requires addressing skepticism, clarifying misconceptions, and fostering a collaborative relationship between scientists and the public.

Xenobot acceptance hinges on mutual understanding. As the "Xenobot Chronicles" unfold, humanity has a unique opportunity to embrace a pioneering field in a way that respects both our ethical boundaries and our shared hopes for a better future.

Analyzing Public Attitudes Toward Xenobots and Biotechnology

As we advance further into the era of biotechnology, the concept of "xenobots" has captured the world's attention, sparking both excitement and apprehension. These tiny, self-replicating biological machines, created by scientists from living frog cells, represent a breakthrough in biotechnology with potential applications in medicine, environmental management, and more.

However, like many groundbreaking technologies, xenobots have also raised ethical, ecological, and safety concerns.

Understanding how the public perceives these novel biological entities is essential to navigating the future of biotechnology responsibly and effectively.

Xenobots are biological robots, constructed from living cells (originally frog embryonic cells) and programmed to carry out simple tasks, such as moving toward a specific target or clustering small objects.

They're named after the African clawed frog species Xenopus laevis, from which their cells are derived.

Designed by combining stem cells with computer algorithms, xenobots have demonstrated unique abilities, such as spontaneous movement, targeted behaviors, and even limited self-replication under specific conditions.

From a technical perspective, xenobots bridge the gap between artificial and natural systems, showing potential applications that range from targeted drug delivery and environmental cleanup to regenerative medicine.

Despite these promising applications, the concept of xenobots has brought forth diverse public opinions, influenced by factors including ethical concerns, potential risks, and the futuristic nature of biotechnology.

Understanding Public Attitudes Toward Biotechnology

Public perceptions of biotechnology have long been complex and nuanced. Historically, reactions to new biological advancements have varied widely, from enthusiastic support to deep skepticism.

Public opinion is shaped by multiple factors, including cultural values, levels of scientific understanding, religious beliefs, ethical considerations, and trust in scientists and institutions.

Xenobots introduce a particularly unique challenge, as they are neither purely "natural" organisms nor traditional "artificial" machines.

This blending of categories challenges traditional ideas about what it means for something to be "alive" or "engineered."

Studies on public attitudes toward biotechnology reveal mixed feelings. People tend to accept technologies they perceive as beneficial and safe, such as vaccines or certain medical treatments, but may resist or question those with unclear or potentially harmful implications, such as genetically modified organisms (GMOs). The newness of xenobots, their capability to self-replicate, and their positioning as both biological and engineered objects mean they often encounter an initial skepticism from the public.

Key Concerns Surrounding Xenobots

Ethical Implications

One of the main ethical questions surrounding xenobots is whether it is morally acceptable to create and manipulate life forms. While xenobots are developed from frog cells and lack a nervous system, questions arise about the boundaries of biological manipulation. Public apprehension is often rooted in fear of "playing God," as creating organisms from scratch or reassembling them from living cells can be perceived as humans encroaching on natural or divine processes.

Some people worry that these technological advancements, while innovative, might eventually cross lines that disrupt fundamental ethical principles regarding life, nature, and human intervention.

Safety and Environmental Impact

Safety is a major concern for the public, especially when dealing with biological entities capable of self-replication. The notion of "runaway xenobots" conjures up dystopian scenarios in which they evolve or proliferate uncontrollably, potentially harming ecosystems. While scientists emphasize that current xenobots are carefully designed to avoid such outcomes, public perception is influenced by these potential risks. The broader concern relates to a fear of unintended consequences, especially in a world where rapid technological advances can sometimes outpace regulations and ethical frameworks.

Potential for Weaponization

Another factor contributing to public unease is the fear that xenobots, like many other forms of biotechnology, could be weaponized. Although xenobots are still in the early stages of development and are not sophisticated enough for such applications, historical experiences with technology being used for harmful purposes contribute to these concerns. The mere possibility that xenobots could one day be adapted for purposes beyond their original intent — potentially even hostile applications — stirs public anxiety about unintended consequences of this powerful technology.

Trust in Science and Government

Public attitudes toward xenobots are also heavily influenced by levels of trust in science and regulatory authorities. When scientific institutions are transparent and communicate openly about risks, benefits, and unknowns, public trust tends to increase.

However, when biotechnology advancements are perceived as shrouded in secrecy or overly technical jargon, they can foster skepticism. Misinformation and sensationalism, often exacerbated by social media, can further erode trust, particularly among groups predisposed to distrust biotechnology. Public outreach and clear communication are therefore critical in helping people form balanced views on xenobots.

Public Reactions: Fear, Fascination, and Hope

Fear and Caution

Many people express concern about the potential dangers of xenobots, from ethical ramifications to environmental risks. The idea of self-replicating biological entities raises fears of unintended ecological consequences, such as altering ecosystems if xenobots were released into the wild.

While these risks are speculative, they resonate with people's intrinsic caution toward self-sustaining technologies, especially when risks are poorly understood or difficult to quantify.

Fascination and Curiosity

On the other side, xenobots have also captured the imagination of many who see them as marvels of science. People are often fascinated by the thought of programmable life forms and what they might mean for the future of technology and medicine.

Xenobots challenge conventional notions about life and technology, inspiring curiosity about what biotechnological advancements might look like in coming years.

Hope for Positive Impact

For some, xenobots represent a source of hope, as their potential benefits align with pressing global needs. Applications in fields such as medicine, pollution cleanup, and environmental management inspire optimism, especially as global crises like climate change and plastic pollution intensify. If xenobots can offer sustainable, effective solutions to these problems, the public may come to view them as valuable tools rather than threats.

The Path Forward:
Public Engagement and Responsible Innovation

As xenobot research progresses, engaging the public in an open and transparent dialogue is crucial. Scientists and policymakers should work together to educate the public about xenobots' benefits, limitations, and safety measures. Addressing people's ethical, safety, and ecological concerns is essential to building trust and understanding.

Equally important is the need for responsible innovation. Developing guidelines and regulations that ensure xenobots are used safely and ethically will be necessary to protect both people and the environment. Public feedback can serve as a valuable resource for policymakers, helping them shape regulations that reflect societal values.

Navigating Public Perceptions for Future Biotechnology

The public's response to xenobots highlights the broader dynamics of public attitudes toward biotechnology — a balance between cautious optimism, ethical debate, and the ever-present need for transparency and responsible action. Xenobots serve as a unique focal point for discussions about the ethical limits of science, the promise of biotechnological innovation, and society's role in guiding responsible technological progress.

As the Xenobot Chronicles continue, it is clear that the path forward requires not only scientific innovation but also careful listening to and understanding of public sentiment.

Strategies for Fostering Positive Public Perception

The emergence of xenobots—a revolutionary innovation in biological robotics—has generated intense interest and curiosity, as well as valid concerns among the public.

These self-healing, programmable organisms made from frog cells represent a significant leap in scientific capability, blending biology and technology in ways never before possible.

As with many scientific advancements, the public's perception is crucial for the responsible and productive development of xenobot technology.

Building a positive perception involves careful, proactive strategies that address public concerns, highlight potential benefits, and ensure transparency.

1. Transparent Communication of Scientific Goals and Limitations

Openly communicating the goals of xenobot research is essential in demystifying the technology. Researchers can begin by clearly explaining the scientific motivations behind developing xenobots, such as understanding cellular self-assembly, developing regenerative medicine, and exploring potential environmental applications like microplastic collection.

By emphasizing these goals, the public can see xenobots as a tool to solve real-world problems rather than as mysterious entities with unknown intentions.

Alongside these goals, it's important to convey the limitations of xenobot technology. Xenobots are not conscious organisms, nor do they possess the capabilities of advanced artificial intelligence.

Outlining these boundaries helps prevent sensationalism, which could lead to exaggerated fears. Educating the public on what xenobots can and cannot do builds trust by giving an accurate picture of their functionality and safety.

2. Promoting Ethical Standards and Safety Protocols

Ethical considerations are a critical part of responsible xenobot development.

Scientists should be transparent about the ethical standards they follow, such as ensuring that xenobots are developed in a way that respects bioethics and avoids harm.

Publicizing safety protocols—such as laboratory containment measures and controlled testing environments—helps to reassure people that researchers are prioritizing the safe handling of xenobots.

Additionally, engaging bioethicists to work alongside researchers can enhance the credibility of xenobot development, as it demonstrates a commitment to moral considerations.

This collaborative approach helps mitigate public concerns about ethical risks and shows that the scientific community is taking proactive steps to prevent any potential misuse or unintended consequences.

3. Involving the Public in the Research Process

Public engagement can be instrumental in building positive perceptions. Researchers can invite community members to participate in discussions or forums, where they can ask questions, voice concerns, and learn about xenobot technology directly from experts.

Public participation events—such as live lab tours, Q&A sessions, and open discussions—create transparency and build a sense of shared ownership over the technology's development.

This kind of engagement is especially effective in addressing concerns that may arise from a lack of information or from misinformation.

When people have the opportunity to learn directly from researchers, they're more likely to feel informed and less likely to believe sensationalized or fear-driven narratives about xenobots.

4. Emphasizing Tangible Benefits for Society and the Environment

Demonstrating the potential benefits of xenobots for society and the environment can help shift public perception in a positive direction. For example, xenobots could play a role in regenerative medicine, offering solutions for wound healing and tissue repair by providing insights into how cells cooperate to form functional structures.

In environmental applications, xenobots could be designed to gather microplastics or other pollutants in oceans, a task that could greatly benefit marine ecosystems.

By focusing on these real-world benefits, researchers can show that xenobots aren't merely a scientific novelty but are aimed at addressing pressing issues.

Emphasizing these applications frames xenobot research as an endeavor that seeks to improve health, sustainability, and the environment, which can resonate with the values and priorities of many people.

5. Utilizing Clear, Accessible Language

Scientific jargon can alienate the general public, making the technology seem more mysterious or intimidating than it actually is. By using clear, accessible language, researchers can break down complex concepts into digestible explanations.

This includes avoiding overly technical terms and instead using analogies or visual aids that make xenobots more relatable. For instance, researchers could describe xenobots as "living tools" rather than "synthetic biological entities" to create a more human-friendly understanding.

Simplified language can also help correct misconceptions. For example, explaining that xenobots are akin to "biological robots" made from frog cells without a brain or consciousness helps people realize that xenobots are controlled by researchers and not autonomous beings with independent thought.

6. Developing a Robust Media Strategy

Media plays a crucial role in shaping public opinion, so it's important for researchers to proactively share accurate, balanced information with journalists and news outlets.

A strategic approach might include offering journalists access to laboratories, providing press releases that emphasize the responsible development and potential benefits of xenobots, and encouraging coverage of real-world applications rather than speculative risks.

Scientists can also participate in interviews, podcasts, and science documentaries to share insights directly with the public, avoiding sensationalism and fear-inducing headlines.

Social media can be particularly effective for reaching younger audiences, and platforms like YouTube or Instagram can be used to share short, informative videos that explain xenobots in an engaging way.

7. Educating the Next Generation

Investing in educational outreach can have a long-lasting impact on public perception.

By introducing xenobot concepts in high school and college curriculums, educators can foster a generation of students who understand and appreciate the science behind xenobots.

Classroom activities, hands-on projects, or even virtual lab experiences that explore cellular biology and robotics help demystify xenobots and encourage informed perspectives.

When students learn about xenobots from an early age, they are more likely to develop a balanced view of the technology, equipped with both enthusiasm for its potential and an understanding of ethical considerations.

This approach builds a foundation of trust and informed curiosity that can positively influence public opinion over time.

8. Demonstrating Responsibility through Regulatory Compliance

Ensuring that xenobot research follows regulatory guidelines and is compliant with local and international safety standards reinforces public trust. Scientists and developers should work closely with regulatory bodies to ensure that xenobots are developed and deployed responsibly. Publicizing this collaboration reassures people that xenobot technology is subject to oversight, thereby reducing fears about "runaway" biological engineering.

By communicating the regulatory safeguards in place, researchers can help the public see that xenobots are being developed within a controlled, responsible framework.

This adherence to regulation helps people feel secure in the knowledge that xenobots aren't advancing unchecked but are subject to scrutiny and accountability.

9. Building Long-Term Trust through Consistency and Accountability

Finally, fostering a positive perception of xenobots is an ongoing process. Maintaining a commitment to transparency, ethics, and open communication builds long-term trust between researchers and the public. Researchers can encourage feedback, report openly on research outcomes—both successes and challenges—and address public concerns as they arise.

This accountability shows that scientists are committed to making xenobot development a collective, responsible effort rather than a secretive pursuit.

These strategies collectively aim to demystify xenobots, address ethical concerns, and emphasize the technology's potential to benefit humanity and the planet.

With consistent and proactive engagement, the scientific community can foster a public perception of xenobots as a responsible, ethical innovation with valuable applications for the future.

Chapter 25: Xenobots Beyond Earth

In the realm of space exploration, xenobots—biological robots crafted from frog cells—are emerging as an innovative tool with transformative potential. As humanity's interest in deep space and planetary exploration grows, scientists are investigating the unique capabilities of xenobots and envisioning their use in environments beyond Earth.

These tiny biological machines, first developed in 2020, have spurred curiosity for their remarkable adaptability, environmental responsiveness, and regenerative abilities, making them a promising candidate for applications that lie well beyond traditional technology.

The Need for Innovation in Space Exploration

Space is a realm of extremes. Mars, the Moon, and even further destinations like Europa, one of Jupiter's icy moons, all possess conditions that are vastly different from Earth's: high levels of radiation, extreme temperatures, reduced gravity, and scarce resources. Traditional mechanical robots and automated spacecraft play an essential role in exploration; however, they face certain limitations.

Mechanical devices are often bulky, difficult to transport, and susceptible to wear and tear in harsh environments. Batteries and electronics deteriorate over time, and repair is a significant challenge without human intervention.

Xenobots, however, bring an exciting new dimension. Since they are biological entities, they do not require batteries, electronics, or metal frames to function. Instead, they derive energy from their own biological components, regenerate naturally, and can adapt to new surroundings.

These qualities may soon lead to a new paradigm where biologically derived robots support and complement traditional machines in extraterrestrial environments.

How Xenobots Work and Why They're Ideal for Space

Xenobots are made by organizing living cells—usually skin and heart muscle cells from the African clawed frog, Xenopus laevis—into tiny structures capable of autonomous movement. The skin cells form a protective layer, while the heart cells contract rhythmically, enabling movement. These bots can self-assemble, regenerate minor injuries, and even work together in swarms, demonstrating a level of "intelligence" that allows them to perform tasks collectively.

In space, these attributes open up novel possibilities. Xenobots could, for instance, be deployed to crawl into tight, difficult-to-reach areas or adapt to unpredictable terrain. Additionally, since they are made of living cells, xenobots may potentially be engineered to derive energy from environmental sources, such as organic material on distant planets or moon surfaces. This adaptability could extend their mission durations without requiring extensive recharging or external fuel.

Potential Applications of Xenobots in Space

1. Regolith Collection and Analysis

One promising application for xenobots is regolith collection on the Moon or Mars. Regolith, the loose rock and dust on the surfaces of these celestial bodies, contains valuable information about planetary formation and composition. Xenobots could gather samples, maneuvering around obstacles and collecting fine particles. Because of their small size and biological nature, xenobots could even operate in tandem with traditional rovers, complementing larger robotic explorers by accessing narrow crevices or fragile areas where mechanical robots might struggle.

2. Self-Healing Structures for Long Missions

Xenobots' regenerative abilities make them candidates for use in creating self-healing materials in space. Traditional materials are vulnerable to wear and micrometeoroid impacts, especially on long-duration missions. By embedding xenobot-like biological units within materials, spacecraft and habitat structures could potentially "heal" minor damage autonomously, maintaining structural integrity over extended periods.

3. Biological Swarms for Large-Scale Mapping

Mapping unexplored territories is vital for future colonization. Swarms of xenobots could cover large surface areas efficiently.

As biological entities, they can change their configurations, communicate with each other via chemical signals, and respond as a group to environmental changes.

By dispatching these swarms, scientists could gain insights into planetary terrain, uncover hidden caverns, or even scout for mineral-rich zones that could one day support extraterrestrial bases.

4. Radiation-Resistant Shielding

One of the biggest challenges for astronauts is protection from cosmic radiation, which is prevalent outside of Earth's atmosphere.

While xenobots themselves are not inherently radiation-resistant, scientists are exploring the potential for engineering xenobots with cells that can repair radiation-induced damage.

These xenobots could act as "living shields," forming protective barriers for sensitive equipment or habitats on lunar or Martian surfaces.

5. Microbial Interaction Studies

The potential discovery of microbial life on Mars or Europa would be one of the most groundbreaking events in science.

Since xenobots are living organisms, they could play a role in exploring microbial interactions on these bodies.

By studying how xenobots respond to extraterrestrial environments, scientists could gain insights into the adaptability and limits of cellular life, possibly even testing for the survival of similar life forms under these conditions.

Xenobots as Bio-Pioneers of Future Space Missions

The adaptability and environmental responsiveness of xenobots make them a fascinating fit for bio-engineered missions in space. For example, xenobots could be released into Martian caves to investigate their structures, testing their ability to navigate, survive, and adapt in low-light, cold conditions.

They could also contribute to a range of pre-colonization tasks by preparing terrain, testing soil quality, and identifying resource-rich areas, all without the need for constant monitoring from Earth.

In space missions beyond Earth, xenobots could even carry specialized payloads, such as sensors for monitoring environmental conditions or DNA sequencers to analyze soil samples.

As space agencies continue to explore the feasibility of Mars and Moon colonization, xenobots may offer an efficient, lightweight, and highly adaptable way to conduct preliminary studies, minimizing risks for future manned missions.

Challenges and Considerations

Despite their exciting potential, there are challenges to deploying xenobots in space. For one, the survival of these biological machines in high-radiation and low-temperature environments is still being researched. Engineering xenobots to withstand these conditions will require genetic modifications or protective designs to help them endure exposure to space.

Additionally, because they are biological, there is a risk of contamination—if xenobots were to encounter indigenous microbes on a distant planet, the results could be unpredictable.

To address these issues, space agencies are exploring containment protocols that would prevent cross-contamination while maximizing the potential of xenobots to study extraterrestrial environments. Research is also underway to understand how xenobot "generations" might evolve in space and whether this could benefit or hinder missions.

The Future of Xenobots in Space Exploration

As scientific advancements continue, xenobots may soon complement traditional technologies, becoming an integral part of humanity's journey beyond Earth.

Their flexibility and self-sustaining nature present unique advantages for the challenges of deep space, making them valuable allies in the quest to understand and eventually inhabit other planets.

By paving the way for self-regenerative, adaptive, and efficient exploration technologies, xenobots represent a forward-thinking approach to expanding humanity's presence in the universe.

Xenobots and Interplanetary Exploration

Charting a New Frontier
In the vast expanse of space, where planets, moons, and asteroids drift millions of miles from Earth, humanity's dream of exploring beyond our solar system is no longer confined to science fiction.

While humans have landed on the Moon and sent probes to the furthest reaches of our galaxy, the idea of sending intelligent, adaptable life forms as our proxies is an emerging possibility.

One innovative answer to the challenges of interplanetary exploration lies in a groundbreaking technology: xenobots.

Xenobots, created from the stem cells of African clawed frogs (Xenopus laevis), represent a fascinating intersection of biology and technology. These tiny, programmable organisms, less than a millimeter in size, have shown potential in many fields, from targeted medicine delivery to environmental clean-up.

However, their unique characteristics also open new frontiers for space exploration. This chapter dives into the potential of xenobots as explorers of interplanetary landscapes, outlining how these biological machines could reshape our approach to understanding and potentially inhabiting other worlds.

Xenobots as Pioneers in Space Environments

Space exploration is an arduous endeavor requiring tools that are durable, adaptive, and energy-efficient. Traditional mechanical robots, while robust, have limitations when it comes to extreme adaptability and self-repair.

Xenobots offer distinct advantages due to their biological origins: they can heal themselves when damaged, adapt to new environments, and even perform coordinated group behaviors. Such characteristics make them ideal for surviving and exploring hostile, unpredictable alien terrains.

In environments where radiation levels, temperature fluctuations, and pressures far exceed those found on Earth, xenobots' organic resilience could be a game-changer. Traditional electronics are vulnerable to cosmic radiation, which can cause short circuits and malfunctions. But, as living biological entities, xenobots might tolerate these conditions better than purely electronic counterparts. This could extend the duration of exploratory missions and increase the reliability of data collection.

Resource Efficiency and Regenerative Abilities

A primary concern in interplanetary missions is resource management. Any mission must consider the limitations of carrying energy sources and the possibility of in-situ resource utilization (ISRU). Xenobots, as living organisms, possess the ability to metabolize organic material. In theory, if provided with a basic nutrient source, they could sustain themselves for extended periods without requiring large energy payloads. This could drastically reduce the weight and cost of missions, as xenobots could potentially utilize biological resources found on other planets or moons.

Additionally, xenobots have shown an impressive ability to regenerate after injury, making them less dependent on repairs or replacements.

Unlike mechanical devices that break down over time, xenobots can repair cellular damage, regaining functionality within hours or days. This regenerative quality makes them particularly suited for prolonged missions in environments where mechanical assistance is limited.

Autonomous Adaptation to Uncharted Terrains

One of the most challenging aspects of exploring distant planets or moons is the unpredictable terrain. Mars, for example, has sand dunes, rocky plains, and mountainous regions, each presenting unique obstacles for rovers and landers. Traditional rovers like NASA's Curiosity or Perseverance are equipped for specific terrains but may struggle in others. Xenobots, however, have an advantage due to their pliable, soft-body structure, which allows them to adapt to irregular surfaces and squeeze through small spaces.

In experimental settings, xenobots have demonstrated the ability to maneuver through narrow channels and avoid obstacles. This adaptive navigation could prove invaluable in unexplored environments where structural flexibility is a necessity.

Imagine sending a swarm of xenobots into a Martian cave system to explore its intricate tunnels and transmit valuable data back to Earth. Their capability to move independently or in coordinated swarms could also allow them to cover large areas efficiently, mapping alien landscapes with remarkable precision.

Environmental Sensing and Data Transmission

A crucial part of any exploration mission is collecting and transmitting data. Xenobots could be programmed to serve as bio-sensors, detecting chemical or physical changes in their surroundings. Scientists could design xenobots to seek out specific compounds, such as water, methane, or other indicators of possible life. With their sensitivity to environmental stimuli, xenobots could autonomously identify and assess zones of interest, such as mineral deposits or organic material.

One envisioned advancement is equipping xenobots with micro-sensors capable of gathering data on temperature, atmospheric composition, and radiation levels. These sensors could transmit data back to a central receiver, which would then relay information to Earth.

By deploying thousands of xenobots across vast areas, scientists could receive detailed maps and environmental assessments of alien terrains, allowing mission planners to prioritize areas for further investigation.

Potential for Terraforming and Habitat Creation

Beyond exploration, xenobots hold speculative promise for preparing extraterrestrial environments for human habitation. Although this concept is highly futuristic, the idea of using xenobots as agents of terraforming is gaining traction.

Terraforming refers to altering a planet's environment to make it more Earth-like, potentially enabling human colonization. Xenobots could play a role in this process by manipulating local biological resources or producing compounds necessary for habitat formation.

For instance, if xenobots could be designed to generate oxygen or absorb toxic elements, they might contribute to creating more breathable atmospheres on alien worlds. Another speculative application is the construction of micro-habitats.

Xenobots could serve as bio-bricks, aggregating to form complex structures. These bio-engineered habitats could serve as shelters for future human missions, particularly on planets with harsh climates like Mars or Europa.

Ethical and Regulatory Considerations

The deployment of living organisms, even synthetic ones, in extraterrestrial environments raises ethical questions. Sending xenobots into unspoiled alien ecosystems introduces risks of contamination and unintended ecological impact. Although xenobots are not capable of reproduction, their presence in delicate, potentially life-hosting environments could interfere with the natural composition and data integrity of those areas.

Regulatory frameworks governing interplanetary contamination are already in place, managed by international bodies such as the Committee on Space Research (COSPAR). These guidelines aim to prevent biological contamination of celestial bodies, especially those that might harbor life. Before xenobots can be deployed in space, extensive assessments will be required to ensure compliance with these standards. Scientists will need to establish protocols for xenobot sterilization, containment, and retrieval to minimize the risk of ecological disturbance.

The Future of Xenobots in Space Exploration

While xenobots are still in the early stages of development, their potential applications in space exploration are vast and imaginative. As research progresses, we may see xenobots become an integral part of humanity's toolkit for space exploration. Their ability to adapt, regenerate, and perform tasks autonomously offers an enticing glimpse into a future where biology and technology work hand-in-hand to push the boundaries of what is possible.

Ultimately, the journey of xenobots from Earth to other planets reflects a shift in how we approach extraterrestrial exploration. By harnessing the resilience of biology in the extreme conditions of space, xenobots may pave the way for a new era in which life itself becomes our most valuable explorer. In the quest for knowledge beyond our world, these remarkable bio-machines could prove to be our greatest allies, transforming the dream of interplanetary exploration into a reality.

Xenobots and the Future of Space Exploration

Imagine a new age of space exploration where autonomous, self-repairing, and even self-replicating biological machines travel alongside astronauts, performing critical tasks to ensure human survival in the harshest environments. This vision is closer to reality than we might think, thanks to advances in biotechnology that have led to the development of xenobots—living, programmable micro-organisms capable of performing specific functions.

These millimeter-sized, mobile constructs are not merely curious lab creations but hold the potential to transform our approach to space exploration and establish a human presence on other planets.

Xenobots are created from the cells of the African clawed frog (Xenopus laevis), giving them a unique biological foundation.

Unlike synthetic robots, xenobots are made from living tissue, specifically repurposed stem cells that have been assembled in specific patterns to enable movement, response to environmental cues, and even cellular repair.

These cells are capable of surviving in hostile conditions, maintaining life-supporting activities autonomously, and reproducing by forming new organisms—a characteristic that is particularly fascinating for researchers exploring their applications in space.

The potential of xenobots lies in their adaptability, biocompatibility, and efficiency. Unlike traditional robots that require complex machinery and power sources, xenobots rely on the inherent energy within living cells, simplifying their design and significantly reducing the need for onboard power. Their biological foundation allows them to degrade naturally when their mission is complete, making them ideal for long-duration missions where minimizing waste is crucial.

Why Xenobots Are Ideal for Space

Space is a hostile environment with extreme temperatures, high levels of radiation, and limited resources. To establish a human presence on the Moon, Mars, or other celestial bodies, we need technologies that can withstand such conditions, self-repair, and work autonomously. Xenobots are uniquely suited for this purpose due to their regenerative properties and autonomous function. They don't require constant human oversight, can adapt to new situations, and can heal minor damage on their own—qualities that are invaluable for exploration far from Earth.

Additionally, xenobots can survive on minimal resources and exhibit efficient locomotion, making them capable of conducting scouting missions on alien terrain or probing environments unsuitable for humans. For example, a fleet of xenobots could be deployed to explore the surface of Mars, mapping its terrain, and collecting samples from areas that may be too risky for human astronauts.

Practical Applications of Xenobots in Space

1. Resource Collection and Processing

A crucial aspect of human space exploration is the ability to find and process resources for survival and habitat construction. Xenobots could assist in the collection of basic resources such as water and mineral deposits.

They could be engineered to identify and gather essential resources by autonomously seeking specific environmental cues, such as moisture or certain mineral compounds.

In extraterrestrial environments, xenobots might perform "biological mining"—extracting essential compounds from the soil to make them usable for humans. Imagine a team of xenobots gathering Martian soil and processing it for water or oxygen, significantly reducing the amount of resources needed to be transported from Earth.

2. Autonomous Repair and Maintenance

Maintaining a base on another planet will inevitably involve wear and tear due to the extreme conditions. Xenobots could be used to autonomously repair minor damage to structures, helping ensure that habitats and critical equipment remain functional without constant human intervention. Their regenerative capabilities are particularly useful here, as they could potentially heal small cracks in a habitat wall or reassemble themselves after damage, reducing the reliance on traditional maintenance methods that require costly and bulky spare parts.

Furthermore, xenobots could be engineered to form biofilms or self-assembling structures that strengthen and protect key areas of a habitat, mitigating damage from radiation or micrometeorites over time.

3. Medical Applications for Astronauts

Long-term space missions pose significant health risks due to isolation, limited medical resources, and prolonged exposure to cosmic radiation. Xenobots could be developed to monitor astronauts' health, deliver targeted drugs, or even perform microsurgery. Imagine tiny xenobots circulating within an astronaut's body, identifying and repairing damaged cells, or delivering medication precisely where it is needed.

In addition, xenobots could aid in immune support, protecting astronauts from microbial threats by directly targeting and neutralizing any foreign pathogens they might encounter during their journey or in a space colony.

4. Environmental Adaptation

Establishing a self-sustaining presence in space requires creating an environment that can support life. Xenobots could be used to create bio-habitats by preparing soil, seeding microbial life, or generating nutrients necessary for the growth of other organisms. By preparing these bio-habitats, xenobots could help establish a foundation for creating biodomes or enclosed ecosystems that recycle oxygen, manage waste, and produce food.

For instance, xenobots could be programmed to break down waste materials into compost or other usable forms that contribute to a closed-loop life-support system. Over time, this could lead to a more sustainable method of supporting human life on other planets.

Challenges and Future Research

Although the potential applications of xenobots in space are groundbreaking, several challenges remain. The durability of living cells in extreme conditions, such as high radiation levels and microgravity, is still under investigation.

Researchers are exploring methods to improve the resilience of xenobots by incorporating genes or cellular mechanisms from extremophiles—organisms that thrive in extreme conditions.

If successful, this bioengineering could yield xenobots capable of functioning optimally even in the most inhospitable environments of space.

Another key area of research is ensuring the ethical and environmental impact of xenobot deployment. As autonomous, living entities, xenobots raise ethical questions about control and environmental effects, especially in unexplored ecosystems.

Scientists are working on stringent programming and control measures to ensure that xenobots can be fully neutralized or naturally biodegrade when their mission ends, avoiding unintended consequences in extraterrestrial ecosystems.

The Future: A New Era of Space Exploration

Xenobots represent an innovative approach that leverages the synergy between biology and technology, combining the adaptability of living organisms with the precision of robotics. Their role in establishing a human presence in space could be transformative, enabling autonomous resource collection, habitat maintenance, and even healthcare support.

As our understanding of xenobots deepens and technology advances, these small biological machines may one day pave the way for sustainable human settlements beyond Earth.

The age of xenobots is still in its infancy, but their potential in the context of space exploration is profound. In the coming decades, xenobots could become indispensable members of space exploration teams, silently working alongside astronauts, ensuring that humanity can explore, settle, and thrive in the vast, uncharted frontier of space.

Chapter 26: The Evolution of Xenobot Diversity

Scientists have made a remarkable leap in synthetic biology with the development of xenobots, engineered biological machines crafted from living cells. These tiny organisms, often referred to as "living robots," represent a groundbreaking synthesis of technology and biology, opening doors to a new kind of bioengineering that merges scientific innovation with nature's own materials. Initially limited in function, early xenobots were capable of simple, programmed behaviors.

However, as the science has progressed, these organisms have diversified both in form and function, offering a glimpse into an exciting frontier where xenobot diversity could address complex problems in medicine, environmental science, and even space exploration.

This chapter explores how the diversity of xenobots has evolved, examining the types of designs, control mechanisms, and potential applications that have emerged as researchers push the boundaries of what these biological robots can do.

The First Generation: Simple Forms with Basic Functions

The initial generation of xenobots was designed by a team of biologists and computer scientists at the University of Vermont and Tufts University, using frog embryo cells to create a basic structure capable of movement. These early xenobots were crafted by arranging clumps of skin and cardiac cells. The cardiac cells provided the beating rhythm needed for locomotion, while the skin cells offered a stable scaffold, creating a simple organism capable of self-propulsion.

Programmed to move in a predetermined direction or cluster in groups, these first xenobots demonstrated controlled behavior without a nervous system or sensory organs. Although limited in their abilities, they served as a crucial proof of concept, showcasing that cells could be reconfigured into entirely new life forms with programmed behaviors.

Expanding Form and Function: A New Level of Complexity

After proving that single-purpose xenobots could be made to move predictably, researchers began experimenting with variations in shape, cell composition, and function. By altering the arrangement of skin and heart cells, xenobots could now navigate around obstacles, pick up small particles, or even work collaboratively to move objects.

The diversity in their designs reflected growing sophistication in their construction. For instance, xenobots were now designed in shapes such as rings, cubes, and even star-like structures, each form enhancing different abilities, like increased movement efficiency or greater durability.

Advances in AI-assisted design also played a pivotal role in this evolution. By using algorithms to simulate millions of potential shapes and configurations, scientists could predict which forms would perform specific tasks most effectively. In this way, computational biology has enabled researchers to create an increasingly diverse range of xenobots tailored to specific tasks. These advancements led to the production of "task-specific" xenobots, with some optimized for targeted drug delivery and others designed for environmental cleanup, such as collecting microplastics in water.

Autonomous Adaptability:
Xenobots with Self-Healing and Reproductive Capabilities

One of the most surprising discoveries in xenobot evolution was the realization that these tiny organisms could exhibit self-healing capabilities. Researchers found that when cut or damaged, xenobots could heal themselves and continue functioning, making them particularly resilient.

This adaptation came as an unexpected advantage in the pursuit of more durable, autonomous bio-machines. Self-healing xenobots could have potential applications in challenging environments where repairs might be difficult or impossible, such as in deep-sea exploration or contaminated areas.

Even more astonishing was the observation that xenobots could "reproduce" under certain conditions. This form of reproduction isn't traditional; rather, xenobots can gather free-floating cells in their environment to create new generations. By aggregating these cells into clusters that eventually develop into fully functional xenobots, this phenomenon, known as "kinematic replication," has pushed the boundaries of what researchers previously thought possible.

Though still in the early stages, the development of self-replicating xenobots has sparked significant interest, as they could lead to sustainable, self-perpetuating systems capable of performing tasks over extended periods without human intervention.

Integrating Sensing and Response:
Towards Intelligent Xenobots

To advance xenobot diversity, researchers have begun exploring ways to incorporate sensory capabilities, allowing xenobots to respond autonomously to their environment. This breakthrough would mark a significant shift from passive to active organisms that can detect stimuli like temperature, light, or chemical presence and adjust their behavior accordingly.

By introducing genes responsible for sensory responses into xenobot cellular structures, scientists are testing prototypes that can navigate environments more intelligently.

For example, xenobots designed for medical applications could be programmed to identify specific cellular markers associated with disease. These xenobots could then deliver medication directly to targeted areas, reducing side effects and enhancing treatment efficacy.

Environmental xenobots, meanwhile, could be designed to detect pollutants or toxins, acting as mobile biosensors in oceans, rivers, or even within city infrastructure.

This expansion into sensing and response represents a leap toward creating xenobots that can act independently and purposefully, functioning almost like "smart" biological agents.

Towards a Future of Cooperative Xenobots

As xenobot designs become more specialized, researchers are investigating ways to enable them to work in teams, mimicking cooperative behaviors found in social animals.

Using swarm intelligence models, scientists are testing how groups of xenobots can coordinate to achieve collective goals.

This cooperation could make them significantly more effective for tasks such as clearing debris, monitoring environmental changes, or even facilitating tissue regeneration in medical settings.

The cooperative aspect of xenobots is being tested by programming them to respond to each other's movements, creating synchronized behaviors.

This group dynamic could be especially valuable in the field of regenerative medicine, where xenobots working in teams might aid in rebuilding damaged tissues or organs.

By embedding communication signals within the cellular structure, these future xenobot swarms could perform intricate tasks in unison, an ability that might lead to transformative applications in both human health and environmental sustainability.

Evolution by Design:
The Road Ahead for Xenobot Diversity

The evolution of xenobot diversity has only just begun, and the future possibilities are vast.

Researchers are exploring the potential to incorporate neural tissue into xenobot structures, which would allow them to process information more complexly and, potentially, even "learn" from their environment.

This would transform xenobots from simple bio-machines into adaptive organisms capable of making autonomous decisions.

As xenobot diversity expands, ethical and regulatory considerations become crucial. Xenobots are fundamentally different from traditional robots or engineered organisms because they exist at the intersection of life and technology.

This unique position raises questions about safety, environmental impact, and the moral implications of creating semi-living, semi-engineered organisms.

Balancing these ethical considerations with the exciting potential of xenobot diversity will be key to guiding future research responsibly.

From their origins as simple clusters of frog cells with limited movement, xenobots have evolved into increasingly complex organisms with a growing array of abilities. This journey of evolution highlights the power of interdisciplinary research, where computational design, cellular biology, and engineering merge to create a new class of bio-machines.

As the diversity of xenobots continues to expand, these remarkable organisms promise to reshape fields from medicine to environmental science, heralding a new era of biologically integrated technology. The evolution of xenobot diversity serves as a testament to human ingenuity and a glimpse into the profound impact bioengineering can have on the world.

Creating Xenobot Ecosystems with Diverse Species and Roles

Xenobot ecosystems, although still in their conceptual and experimental stages, represent an intriguing future possibility where self-organizing systems of biologically engineered organisms can exist, interact, and even adapt. These ecosystems aim to replicate the diversity of natural biological environments by introducing multiple "species" of xenobots, each with distinct roles and capacities.

By developing systems where different types of xenobots can interact harmoniously, scientists hope to mimic complex biological communities, creating a network of organisms that could one day perform essential functions in medical, environmental, and technological settings.

Conceptualizing Xenobot Species and Their Roles

Creating a xenobot ecosystem begins with the idea that diverse roles are necessary for a balanced and functional environment. This mirrors the principles of natural ecosystems, where biodiversity supports resilience and adaptability.

In this model, different "species" of xenobots are designed with specialized purposes, such as nutrient cycling, pollutant removal, communication, and structural support.

For instance, smaller xenobots might serve as "scavengers" by consuming cellular debris or unwanted materials, while larger xenobots might act as "builders," creating small physical structures or even aiding in tissue repair within the body.

The diversity of xenobot functions requires a modular approach to design. By altering the size, shape, and programmed behaviors of xenobots, scientists can engineer unique capabilities tailored to specific roles.

Additionally, genetic or biochemical alterations might allow different types of xenobots to interact selectively or even communicate through chemical signals, adding layers of complexity to their community behavior.

Designing Cooperative Interactions: Communication and Signal Processing

Communication is fundamental to creating a functional xenobot ecosystem. For example, in natural ecosystems, organisms exchange information through chemical signals, vibrations, or even visual cues.

Similarly, xenobots could be designed to communicate using biochemical signals, potentially based on fluorescent markers or electrical impulses.

By sensing these signals, xenobots could coordinate their actions to respond to environmental stimuli.

In a hospital environment, for instance, xenobots programmed to recognize infection markers could summon "healer" xenobots, which might release targeted antibiotics or anti-inflammatory compounds. Other xenobots could function as "scouts," detecting changes in the ecosystem or signaling to nearby xenobots when an area requires maintenance.

These roles could be especially useful in wound healing, where coordinated actions among xenobot species might accelerate recovery by efficiently clearing damaged cells, delivering drugs, and fostering new tissue growth.

Creating signal-processing systems in xenobots would require advances in synthetic biology and bioengineering. Researchers are experimenting with cellular pathways that allow cells to detect, interpret, and respond to specific stimuli, mimicking sensory functions in natural organisms.

With improvements, this communication layer could make xenobot ecosystems increasingly responsive to environmental conditions and capable of cooperative decision-making.

Resource Management and Self-Sustaining Xenobot Ecosystems

For a xenobot ecosystem to be self-sustaining, resources such as nutrients, energy, and waste management need to be balanced. Engineers might design certain types of xenobots to recycle biological waste, converting it into forms of energy or nutrients that other xenobots can use. This could be achieved by engineering xenobots with metabolic pathways similar to bacteria that break down organic materials, ensuring that waste products are efficiently recycled within the system.

Energy management is another critical component of self-sustaining xenobot ecosystems. Current xenobots are primarily powered by stored embryonic stem cell energy, which is limited. In future ecosystems, researchers are exploring the idea of xenobots that derive energy from their surroundings, perhaps by photosynthesis or by metabolizing nutrients from the environment, like algae or small plants. By introducing energy-harvesting xenobots, scientists could create a renewable energy cycle within the ecosystem, enhancing its sustainability and longevity.

Dynamic Adaptability:
Evolving Xenobot Populations

Adaptability is one of the most impressive attributes of natural ecosystems. While today's xenobots are designed with specific, fixed behaviors, future xenobot ecosystems could incorporate genetic algorithms or machine learning capabilities that allow these artificial organisms to "learn" and adapt over time.

By integrating adaptive algorithms, researchers might enable xenobot populations to respond to challenges or changes in their environments in a way that mirrors biological evolution.

For instance, if a specific xenobot task is frequently required—like removing certain toxins—the ecosystem could be designed to recognize this need and "evolve" more toxin-cleaning xenobots.

These changes wouldn't involve true genetic evolution but could instead rely on mechanisms of programmable responses, where the xenobots receive information on task importance and adjust their operations accordingly.

Safeguards and Ethical Considerations

As xenobot ecosystems move from conceptual designs to experimental implementations, ethical considerations and safety measures will be crucial. Researchers must ensure that these self-sustaining systems remain confined and controlled to prevent unintended spread or interactions with natural ecosystems. Xenobot systems could be equipped with "kill switches" or programmed life spans to ensure they deactivate under specific conditions, reducing the risk of environmental contamination or unintended mutations.

Ethical questions arise regarding the autonomy and programming of these synthetic organisms. While they lack consciousness or independent thought, the increasing complexity of xenobot systems raises questions about the potential for unintended behaviors or interactions. Researchers are considering regulatory frameworks to guide the development and use of xenobot ecosystems, ensuring that these systems benefit humanity without posing unforeseen risks.

Future Applications of Xenobot Ecosystems

The potential applications of xenobot ecosystems are vast, spanning medical, environmental, and industrial domains. In medical settings, xenobot ecosystems could be deployed to maintain or repair specific tissues in the human body.

Imagine a xenobot ecosystem that functions like a "healing garden" within a patient's body, cleaning up waste products, delivering drugs, and assisting in wound repair.

This could revolutionize healthcare by providing continuous, in-body care that adjusts to a patient's changing conditions.

Environmental applications are equally exciting. Xenobot ecosystems could be designed to manage waste in water sources or to help maintain soil health. In polluted environments, these systems could work continuously to break down harmful compounds, providing a sustainable, biologically integrated solution to environmental restoration.

Industrially, xenobot ecosystems could be engineered to maintain equipment, detect damage, or even manage micro-environments in space exploration habitats.

Toward a New Symbiosis

Creating xenobot ecosystems is about more than simply engineering new biological entities—it's about fostering a new type of symbiosis between artificial and natural worlds.

As we advance in designing and deploying these ecosystems, we may find that xenobot communities could enhance our capacity to manage and care for both human health and environmental systems in profound ways.

Though many challenges remain, the future holds the promise of ecosystems that are not only adaptive but also truly integrated, bridging the gap between the biological and technological realms in ways previously unimaginable.

Exploring the Potential Benefits of Xenobot Biodiversity

Imagine a future where biology and technology combine to create a world of microscopic robots that can carry out tasks beyond our wildest dreams, and perhaps even beyond human ability. In the field of synthetic biology, xenobots stand as a revolutionary advancement.

These tiny, living robots, crafted from frog stem cells and named after the African clawed frog (Xenopus laevis), are not just scientific novelties—they may become transformative tools across numerous fields.

Biodiversity among xenobots could amplify their utility and sustainability in ways that parallel the advantages of biological diversity in nature.

This concept of xenobot biodiversity holds promise for enhancing the functionality, resilience, and adaptability of these bio-hybrid organisms, potentially offering countless benefits for fields like medicine, environmental conservation, and bioengineering.

1. Expanding Functional Range and Adaptability

Just as biodiversity in nature fosters ecosystems resilient to change, diverse xenobot designs could lead to a suite of bio-robots that can perform distinct and complementary functions. By creating xenobots with varied shapes, structures, and behaviors, scientists can tailor them to meet a wider range of operational needs.

For instance, some xenobots might be specialized to seek and bind to specific pollutants, while others could deliver precise therapeutic agents in the human body.

Imagine a team of xenobots engineered with unique features—some optimized for high-speed movement, others for carrying heavier loads, and yet others for detecting particular chemical markers.

Such diversity would allow xenobots to adapt to specific environmental or medical conditions, maximizing their utility in dynamic and complex settings.

2. Enhancing Resilience and Stability

Biodiversity offers natural systems a safeguard against sudden collapse by providing genetic variations that can help populations survive challenges. A similar approach can be applied to xenobot design.

Different xenobot variants, with unique physical and functional characteristics, could be more resilient in diverse environments, from extreme temperatures to varying chemical conditions.

This diversity would minimize the risks of catastrophic failures in applications such as environmental clean-up or internal human treatments, where xenobots may encounter fluctuating or hostile environments.

By programming distinct behaviors and physical properties into various xenobot types, scientists can ensure that if one group becomes less effective, others will continue functioning, making the whole system more robust.

3. Synergistic Teamwork in Complex Tasks

In natural ecosystems, biodiversity enables species to collaborate in fulfilling ecosystem functions—plants provide oxygen, herbivores control plant growth, and predators maintain population balance. Xenobot biodiversity could enable similar collaborative benefits. Different xenobot types could interact and work together to achieve goals that single-purpose xenobots might struggle with alone.

For example, in environmental clean-up efforts, certain xenobots might specialize in locating pollutants, while others break them down or transport them.

In the human body, some xenobots could navigate to infected areas, while others could deliver medicine directly to targeted cells.

This kind of coordinated teamwork across different xenobot "species" could lead to much more effective outcomes, with each type performing its designated role within a larger operational framework.

4. Accelerated Problem-Solving Through Evolutionary Algorithms

Xenobot biodiversity also has the potential to benefit from evolutionary algorithms, which mimic natural selection.

By experimenting with different xenobot designs, scientists could encourage survival-based variations that improve efficiency and adaptability over generations.

Diverse xenobot designs with various strengths could be created, and those that best perform their intended function would "survive" for replication and further testing.

Over time, this process would yield a diverse pool of optimized xenobots suited to different tasks and conditions. Such an evolutionary approach could rapidly accelerate the discovery of optimal xenobot configurations for specific applications, much like how nature's diversity has allowed organisms to adapt to a wide range of environments over millennia.

5. Increased Efficiency in Resource Utilization
In many industries, efficiency is limited by a "one-size-fits-all" approach, where a single design is adapted to meet multiple needs.

Xenobot biodiversity, however, could enable a more targeted use of resources by deploying specialized xenobots for specific tasks, thus reducing waste and maximizing output.

In agriculture, for example, specialized xenobots could be used for pest control, pollination, or nutrient delivery, each designed with unique attributes suited to their respective roles.

This would lead to more sustainable practices, as resources would be used more efficiently.

Similarly, in wastewater treatment, distinct xenobot types could handle different pollutants, allowing for precise and effective remediation without excessive consumption of resources.

6. Reducing Environmental Impact and Supporting Eco-Friendly Solutions

One of the most promising aspects of xenobot biodiversity is its potential to support eco-friendly solutions in fields that often rely on harsh chemicals or invasive techniques.

Diverse xenobots, each tailored to specific tasks, could offer gentler alternatives to chemical treatments in areas like agriculture, pollution management, and even human health.

Xenobots with biodegradable properties could be designed to naturally break down after completing their tasks, reducing the ecological footprint and lessening environmental impact.

For instance, xenobots that degrade into harmless organic matter could be used in cleaning oil spills or other hazardous waste without leaving behind toxic byproducts, ensuring a more sustainable environmental intervention.

7. Expanded Potential for Personalized Medicine

In the medical field, xenobot biodiversity could revolutionize personalized medicine. Just as human physiology is complex and diverse, so too could xenobot "populations" be diversified to treat individual patients based on unique biological markers or health conditions.

Xenobots could be developed to recognize specific genetic markers or cell types, allowing them to deliver targeted treatments to the correct areas with remarkable precision.

For example, a patient with a particular tumor type could be treated with xenobots designed to recognize and target that specific tumor, minimizing harm to surrounding tissues.

This approach could lead to highly effective treatments tailored to the specific needs of each patient, paving the way for more customized and responsive healthcare solutions.

8. Enabling Deeper Scientific Insight into Bioengineering

Finally, exploring xenobot biodiversity can drive greater scientific understanding and innovation. Each new xenobot variant presents a unique opportunity to study the interactions between biology and technology, shedding light on how life forms can be engineered for optimal performance in synthetic systems.

By developing a diverse array of xenobot types, scientists can observe which designs work best for specific challenges, thereby gaining insights into fundamental biological and bioengineering principles.

This ongoing exploration of xenobot biodiversity could lead to groundbreaking discoveries in areas such as cell differentiation, regenerative biology, and hybrid organism behavior, ultimately deepening our understanding of both biology and bioengineering.

In conclusion, xenobot biodiversity holds enormous potential for addressing some of today's most pressing challenges. By embracing diversity in design and function, researchers could develop xenobots that are more versatile, resilient, and capable of complex tasks.

As this field continues to evolve, xenobot biodiversity may become a crucial asset in fields ranging from medicine to environmental science, creating possibilities limited only by our imagination and ingenuity.

Chapter 27: Xenobots and Neurological Research

In the evolving landscape of neurological research, xenobots represent a fascinating frontier. These microscopic, self-assembling biological robots are created from living cells, typically harvested from the African clawed frog (Xenopus laevis).

By arranging frog skin and heart cells into specific shapes, scientists have engineered these "living machines" to move, respond to stimuli, and even self-replicate.

Their behavior, adaptability, and cell-based design have sparked growing interest in their potential applications within neurological research, particularly in studying brain functions, cellular communication, and even potential therapeutic uses.

The Link Between Xenobots and Neurology

Neurology is a field deeply concerned with the organization, structure, and function of cells and networks within the brain and nervous system.

Traditional models in neurological research rely on simplified in vitro systems or animal models, which often fail to capture the full complexity of human neuronal interactions.

Xenobots, by contrast, can serve as customizable, controlled models that respond to external inputs in real-time, offering a unique way to examine cell signaling, interaction, and response in a biologically active setting.

Although xenobots themselves do not possess neurons, their cell-based construction and ability to exhibit collective behavior are highly relevant to understanding cell dynamics in neurological contexts.

Cellular Communication and Signal Transmission

Understanding how cells communicate is central to neurology. The nervous system relies on intricate chemical and electrical signals to transmit information, control movement, and process thoughts and emotions.

Xenobots have shown the potential to help study these signaling processes at a cellular level. By introducing neurotransmitters or other signaling molecules to a xenobot environment, researchers can observe how these cell clusters react, adapt, or change direction.

This controlled setting could lead to insights on how different compounds influence cellular behavior, mimicking how neurons respond to neurotransmitters.

Furthermore, xenobots have shown a degree of environmental awareness—they can navigate through spaces, avoid obstacles, and group together. By engineering xenobots to respond to electrical or chemical cues, researchers can explore cellular signaling and examine parallels with how neurons communicate across synapses.

Through such experiments, scientists could analyze how neural-like behavior emerges from non-neuronal cells and gain a better understanding of how basic signaling may have evolved in primitive nervous systems.

Modeling Neural Network Dynamics

In neurological research, neural networks—interconnected clusters of neurons that coordinate various functions—are essential for studying memory, cognition, and sensory processing. Xenobots, with their programmable collective behavior, can be used to create simplified neural network models that mimic certain properties of neuron clusters.

When assembled in particular configurations, xenobots can be influenced by external stimuli, enabling them to "learn" paths or exhibit memory-like behavior by adapting their movement based on previous encounters.

While these actions are primitive compared to actual neural networks, they provide a controllable and customizable model for studying networked behaviors.

In the future, researchers could incorporate neuron-like cells within xenobots to create living, responsive networks that operate similarly to artificial neural networks, allowing neuroscientists to test hypotheses about information processing in a simple biological model.

This could deepen our understanding of how neural circuits operate and provide insights into the dynamics of cognitive functions.

Xenobots as Tools for Brain Injury and Disease Modeling

Xenobots offer exciting possibilities for studying neurodegenerative diseases and brain injuries. These conditions often involve disruptions in cell communication, structural changes, and the degeneration of neuronal tissue.

By experimenting with xenobots under various stressors—such as exposure to neurotoxic chemicals, changes in temperature, or oxygen deprivation—scientists can study how cellular damage occurs and progresses.

Observing these responses could serve as a microcosm for understanding certain neurodegenerative diseases, such as Alzheimer's or Parkinson's, where cellular resilience and response to damage play critical roles.

Additionally, xenobots could serve as bio-compatible testing platforms for new drugs aimed at preventing or repairing neural damage. Unlike traditional cell cultures, xenobots present a more dynamic model that moves, responds, and interacts with its environment, allowing researchers to better assess how potential treatments might work in an active cellular network.

This kind of research could accelerate the testing and discovery of new compounds that might protect or regenerate neurons in a more realistic, albeit simplified, environment.

Investigating Neural Plasticity and Adaptation

Neuroplasticity—the brain's ability to adapt and reorganize itself—has been one of the most exciting areas of modern neurological research.

Although xenobots do not possess a nervous system, they do exhibit forms of "adaptation" through their programmed responses to stimuli and even some degree of autonomous behavior, especially in response to environmental challenges.

By exposing xenobots to diverse conditions, scientists can study how cell-based systems adapt and potentially reorganize themselves, offering insights that could correlate to neural plasticity.

In future experiments, researchers may try to integrate neural stem cells or genetically modified cells with neuron-like properties into xenobot structures. This could potentially create "neurobots" capable of reacting in more complex ways, perhaps even showing rudimentary forms of memory or adaptation.

Such advances would allow scientists to probe fundamental questions about how plasticity arises, how cells adapt to new conditions, and how these processes might be harnessed in therapeutic interventions for neurological disorders.

Toward Future Therapies: Xenobots as Neurological Repair Agents

One of the most ambitious applications of xenobots in neurology lies in their potential as repair agents for the nervous system.

Because xenobots are constructed from living cells, they are inherently biocompatible and could theoretically interact with human tissues without eliciting an immune response.

In the future, xenobots could be designed to deliver neuroprotective agents, clear damaged cells, or even create supportive cellular structures in regions affected by injury or disease.

Imagine a xenobot programmed to navigate toward damaged neural tissue and release growth factors that encourage repair or reduce inflammation. Alternatively, xenobots could be developed to bridge small gaps in damaged neuronal circuits, providing temporary scaffolding that would allow neurons to regenerate.

Although these ideas are still speculative, they illustrate the potential of xenobots to serve as adaptable, intelligent therapeutic tools capable of addressing neurological damage in ways that traditional treatments cannot.

Ethical and Safety Considerations

As with any emerging technology, the use of xenobots in neurological research raises ethical and safety questions. While xenobots are cell-based machines and lack consciousness, their use in studying complex neurological processes brings up concerns about manipulation of living cells and unintended consequences.

Regulatory oversight will be crucial to ensuring safe development, especially if xenobots are ever introduced into clinical settings.

Furthermore, as xenobot technology progresses, scientists must carefully monitor how these bio-hybrids behave over time, ensuring they remain controllable and non-invasive in both laboratory and potentially clinical environments.

Addressing these ethical considerations transparently will be essential for public acceptance and the responsible use of xenobots in advancing neurological research.

Xenobots open up a frontier of possibilities in neurological research, bridging cellular engineering and neural science in groundbreaking ways. From modeling neural networks and disease progression to potentially aiding in neural repair, xenobots are redefining how scientists approach the study of the nervous system. As this technology advances, it holds the promise of not only expanding our knowledge of brain function but also developing new methods for treating neurological disorders, bringing the vision of bioengineered neural therapies closer to reality.

Using Xenobots to Study Brain Development and Neurological Disorders

The field of neuroscience is deeply focused on understanding the intricate workings of the brain, yet there remain many mysteries surrounding brain development, behavior, and neurological disorders. In recent years, xenobots have emerged as a groundbreaking technology with potential applications far beyond their initial conception. These tiny, bioengineered organisms, developed from frog cells, were originally designed to explore movement and basic functional responses.

However, their adaptability and controllability have led scientists to investigate how xenobots could be used as models to study complex neurological phenomena.

1. Xenobots as a Model for Brain Cell Interaction

One of the most remarkable aspects of xenobots is that they are constructed from living cells, specifically from Xenopus laevis, the African clawed frog. While they don't have traditional nervous systems, they are made from cells capable of interacting and communicating in ways that mimic early cell behavior during development. This unique quality allows researchers to explore fundamental questions about cellular interactions in early brain formation.

Xenobots provide a simpler, more controllable platform where scientists can observe how cells signal and organize themselves. They enable direct observation of cell behavior, growth, and differentiation, which are crucial processes in understanding how brain cells communicate and assemble in the early stages of development.

Because xenobots are built from real, living cells, they offer a natural bridge between purely synthetic models (such as neural networks on chips) and complex, fully developed brains. By manipulating the cells in a xenobot, scientists can study how changes in cellular structure and organization influence neural communication. Insights gathered from these interactions may lead to breakthroughs in understanding how brain tissue organizes itself during early embryonic stages—a critical step in identifying disruptions that could lead to neurological disorders.

2. Studying Neuroplasticity and Neural Repair with Xenobots

Neuroplasticity, the brain's remarkable ability to reorganize itself by forming new neural connections, is essential for recovery from injuries and adapting to new information. Xenobots, though not possessing full nervous systems, exhibit surprising adaptive behaviors at the cellular level, adjusting their movements and responses to changes in their environment. Scientists are leveraging this adaptability to explore concepts similar to neuroplasticity.

By modifying the structures and conditions surrounding xenobots, researchers can observe how cellular structures change and adapt—a process that could be likened to the brain's way of rewiring itself in response to damage or learning.

In terms of neural repair, xenobots can serve as models for understanding how cells respond to injuries, potentially shedding light on how to stimulate brain repair mechanisms.

For example, by inflicting controlled injuries on xenobots, researchers can observe how cells respond, reorganize, and self-repair.

These observations may reveal pathways and signals that could be harnessed to support brain healing, offering hope for patients suffering from traumatic brain injuries or degenerative conditions.

3. Simulating Neurodegenerative Diseases

One promising application of xenobots in neurology lies in the simulation of neurodegenerative diseases like Alzheimer's and Parkinson's.

Scientists can introduce specific gene modifications or environmental changes within xenobot cells to mimic the cellular conditions seen in these diseases.

For instance, researchers may add protein aggregates associated with Alzheimer's to observe how xenobot cells respond or deteriorate over time.

By creating models that replicate the cellular environment of neurodegeneration, scientists can monitor disease progression in real-time, uncovering how cellular networks break down and how damage spreads.

Xenobots may also allow researchers to test potential treatments in a more simplified system. By introducing molecules designed to counteract or slow down degenerative processes within these bioengineered organisms, scientists can gauge the effectiveness of new therapies in a controlled environment.

This approach could lead to safer, faster, and more cost-effective methods for drug discovery, as researchers would first identify promising candidates within xenobot models before moving to more complex animal studies or clinical trials.

4. Exploring Synapse Formation and Neural Communication

Understanding how neurons communicate through synapses is fundamental to unlocking the mechanisms of learning, memory, and cognition. While xenobots lack full neural systems, they allow scientists to explore cell-to-cell communication at a primitive level.

By using electrical or chemical stimuli, researchers can induce xenobot cells to interact in ways that are reminiscent of synapse formation, enabling the study of how cells establish connections and transfer signals.

This model can be valuable for understanding synaptic dysfunction, a feature of many neurological conditions, such as epilepsy and autism spectrum disorders.

Xenobots offer a simplified platform to study how cells establish and maintain these connections and how disruptions in synaptic signaling might lead to abnormal behaviors.

While xenobots are not yet capable of fully mimicking the vast complexity of the brain, their use in investigating early stages of synapse formation could reveal key insights into the processes underlying neural connectivity and brain development.

5. Testing the Effects of Neurological Drugs and Therapies

Traditional drug testing for neurological disorders often requires extensive trials on animal models before moving to human testing. Xenobots could speed up this process by providing a simplified biological platform for testing the effects of new drugs.

By exposing xenobots to various compounds or altering their cellular environment, researchers can observe real-time effects on cell behavior, division, and organization.

This could be particularly beneficial for understanding the cellular impacts of new medications for neurological conditions, allowing scientists to screen for both efficacy and potential side effects at an early stage.

Xenobots could also serve as models for exploring non-drug-based treatments, such as electrical stimulation therapies, which are used in treating disorders like epilepsy and depression.

Researchers can apply electrical impulses to xenobot cells to study how they respond and adjust, offering valuable insights into how brain cells might be stimulated to improve function or alleviate symptoms in patients.

6. Ethical and Practical Advantages

Using xenobots as models for studying brain development and neurological disorders offers several ethical and practical advantages over traditional models. Since xenobots are developed from frog cells and lack complex consciousness or pain receptors, they avoid some of the ethical concerns associated with higher-level animal testing.

Their small size and adaptability make them cost-effective and easily controllable, which means large-scale studies can be conducted without the need for significant resources. Moreover, xenobots' unique construction allows scientists to make rapid adjustments to their genetic or cellular makeup, permitting high-throughput testing and faster cycles of experimentation than would be possible with mammalian models.

Xenobots represent an exciting frontier in neurological research, offering new ways to explore brain development, disease mechanisms, and treatment options. Their unique properties allow scientists to investigate cellular behavior, simulate neurodegenerative conditions, and study neural communication in ways that were previously unimaginable. While they cannot replace the full complexity of human and animal brain models, xenobots have the potential to revolutionize early-stage neurological research and accelerate our understanding of how the brain develops, heals, and sometimes deteriorates.

As research in xenobot technology advances, we may find ourselves closer to unraveling the complexities of the human brain and unlocking therapies for neurological disorders that have long eluded science.

Advancements in Understanding Neural Processes

Xenobots, the living, programmable organisms derived from frog stem cells, have revolutionized the intersection of biology, technology, and artificial intelligence. One of the most intriguing aspects of xenobot research has been the exploration of neural processes—the underlying structures and mechanisms that dictate behavior, decision-making, and responsiveness. Understanding these neural elements in xenobots has propelled advancements that not only enhance our control over these tiny organisms but also offer significant insights into neural processes in more complex biological systems, including humans.

Neural Processes: The Basics

Neural processes refer to the intricate network of interactions that occur within and between neurons, the cells responsible for transmitting information in living organisms. These processes allow for complex behaviors, responses to environmental stimuli, and decision-making, often on the microscopic level. In the context of xenobots, understanding neural processes is less about consciousness and more about recognizing how signals can be programmed, transmitted, and received within these bio-bots to achieve specific outcomes. Neural mechanisms within xenobots are not identical to those in typical animal neural networks; rather, they are purposefully simplified, enabling scientists to observe foundational neural activity in a controllable, laboratory-grown organism.

The Emergence of Artificial Neural Circuits in Xenobots

One of the major advancements in neural process research with xenobots has been the development of artificial neural circuits. Unlike the neurons in the human brain, which are incredibly complex and functionally diverse, the neural circuits in xenobots are programmed to respond to specific cues or stimuli.

This design allows researchers to observe neural functions in a stripped-down format, shedding light on fundamental principles such as response signaling, movement coordination, and pattern recognition.

For example, when xenobots are exposed to certain stimuli like chemical gradients, light, or changes in temperature, these artificial circuits allow them to move toward or away from these sources. This controlled setup creates opportunities to study basic neural responses in action.

As a result, xenobot research has broadened our understanding of how artificial neural circuits can drive behavior and environmental adaptation in simple organisms, creating a blueprint for mimicking these processes in other biological systems.

Simulating Neural Plasticity in Xenobots

A breakthrough concept being explored in xenobot research is neural plasticity—the ability of neural networks to change and adapt in response to experiences or environmental factors. In humans, neural plasticity is crucial for learning and memory. In xenobots, scientists are experimenting with ways to simulate a form of plasticity to observe how these bio-bots might adapt to new stimuli over time.

By repeatedly exposing xenobots to certain stimuli and tracking their responses, researchers are observing how consistent exposure may lead to "learned" behavior patterns. This simulation of neural plasticity in xenobots could provide a valuable testing ground for understanding adaptive behaviors in larger, more complex organisms. Moreover, it gives researchers a way to evaluate how neural networks might be reprogrammed to optimize specific functions, whether in simple xenobots or more advanced synthetic organisms.

Influence of AI in Enhancing Neural Understanding

Artificial intelligence (AI) has been a game-changer in xenobot research, particularly in relation to neural processes. Machine learning algorithms are employed to predict and shape the neural pathways that dictate xenobot behavior.

By training these algorithms with massive datasets of xenobot movements and responses, AI can help refine our understanding of neural connections in xenobots. This collaborative interplay between AI and xenobot development provides insights that transcend xenobots alone.

For example, AI has facilitated the mapping of neural response patterns in ways previously unimaginable, such as predicting how xenobots might respond to novel stimuli they've never encountered.

This is akin to "pre-training" the xenobot's neural network, creating more precise behavior prediction models that can later be applied to understanding similar predictive models in human neural networks.

The use of AI has helped scientists quickly identify and target neural pathways in xenobots, effectively accelerating the pace of discovery and application.

Studying Feedback Loops and Self-Regulation Mechanisms

Another remarkable advancement in xenobot neural research is the exploration of feedback loops—mechanisms by which xenobots can sense their environment, interpret stimuli, and adjust behavior accordingly.

This self-regulation allows them to perform tasks with minimal human intervention.

By embedding feedback loops within neural circuits, scientists are providing xenobots with a rudimentary form of self-awareness, where they can detect changes in their surroundings and respond dynamically.

For instance, xenobots equipped with feedback loops have been observed redirecting their movement based on obstacles or chemical gradients in their environment.

This feature, often compared to reflexive responses in animals, gives researchers a fresh perspective on how feedback mechanisms can be engineered in synthetic organisms.

Understanding and replicating self-regulation on a microscopic level in xenobots sets the foundation for replicating such mechanisms in larger, more complex systems, contributing insights into how simple reflexes might evolve into more sophisticated responses over time.

Implications for Neurological and Cognitive Research

The advancements in understanding xenobot neural processes have significant implications for human neurological and cognitive research.

By studying how xenobot neural circuits are designed, controlled, and modified, scientists gain a blueprint for testing hypotheses about the basic building blocks of cognition and memory.

Xenobots provide an ethical, scalable model for testing various theories of neural network interactions and behavioral responses, allowing researchers to experiment with interventions that would be ethically challenging or infeasible in human studies.

The implications for treating neurological disorders are profound as well.

By exploring how neural processes are adapted and regulated in xenobots, researchers can develop strategies to potentially reprogram damaged neural circuits in human patients.

Future xenobot models might even be used to test early interventions for cognitive diseases by simulating neural degeneration and regeneration processes in controlled ways, offering a clearer picture of how targeted treatments might affect neural function.

Future Directions in Neural Process Research with Xenobots

Looking ahead, the exploration of neural processes in xenobots is expected to focus on enhancing neural complexity and improving the accuracy of their behavioral responses.

By advancing xenobot models to have increasingly complex neural circuits and response patterns, researchers can create bio-bots that more closely mimic living organisms.

This evolution could lead to the development of xenobots capable of more sophisticated tasks, including coordinated teamwork, environmental sensing, and even basic forms of "problem-solving."

In the future, xenobot research may reveal new approaches for understanding and manipulating neural processes in the human brain.

With each experiment, scientists move closer to unlocking the full potential of neural networks within and beyond xenobots, ultimately shedding light on one of biology's greatest mysteries: how simple neural processes give rise to the complexities of life.

Chapter 28: The Uncharted Ethical Territory of Xenobots

Xenobots—biological machines made from living cells—have transformed scientific and ethical landscapes, sparking conversations not only about their scientific potential but also about the ethical challenges they present. As living entities created in laboratories from frog cells, xenobots blur the lines between artificial intelligence, biology, and robotics. They represent a groundbreaking fusion of life and technology, a frontier filled with exciting possibilities but also formidable ethical questions. This exploration of the ethical considerations surrounding xenobots aims to provide a balanced, accessible view of both the benefits and the concerns associated with these groundbreaking creations.

Understanding the Basic Ethics of Xenobots

Ethical concerns regarding xenobots stem from their fundamental nature: they are organisms created by human intervention, crafted from the cells of living organisms yet designed to function in ways that transcend typical biological roles.

This unique status poses critical ethical questions. Are xenobots living creatures with their own rights, or are they tools without inherent moral consideration? At what point does the manipulation of biological materials challenge our definitions of life and nature?

These questions set the stage for more nuanced discussions about the rights and limitations of scientists when manipulating life for specific goals.

Xenobots, unlike conventional machines, do not rely on metal or silicon but instead are crafted from biological tissues that have been genetically guided to perform tasks. This biological material responds to its environment, can self-repair to a degree, and operates in ways similar to living cells.

These traits complicate the ethics of xenobot development, bringing to light questions about how we treat life at the cellular level and whether a degree of respect is warranted for these "living machines."

The Potential Benefits and Ethical Justifications

Supporters of xenobot research highlight the many potential benefits that these biological robots could offer, presenting a compelling ethical argument for their continued development. In the medical field, xenobots could one day aid in targeted drug delivery, removing toxins from the bloodstream, or even repairing damaged tissues without the need for invasive surgeries. In environmental science, they could assist in cleaning up pollution, removing microplastics from oceans, or managing toxic waste, offering eco-friendly solutions to some of humanity's greatest challenges.

From this perspective, the development of xenobots could be seen as ethically justified, given the enormous potential to improve human and environmental health. Ethicists argue that if xenobots can reduce suffering and promote well-being for humans and other species, their use may align with broader ethical principles of utilitarianism, which prioritize actions that lead to the greatest good. However, this view rests on a presumption that the benefits of xenobots will outweigh the potential risks, a balance that is not guaranteed given the unknowns surrounding their long-term impact.

Autonomy and the Question of Control

One of the more controversial aspects of xenobot ethics lies in the question of control. Xenobots, while created by humans, can exhibit forms of "self-directed" behavior, responding to environmental stimuli and even displaying simple forms of autonomy. This autonomous behavior, although limited, introduces a host of ethical concerns. For instance, if a xenobot can navigate its environment and respond to stimuli, does it possess a minimal form of consciousness or awareness? While xenobots are not sentient by any scientific measure, their autonomous capacities challenge existing ethical frameworks for non-sentient machines and demand a closer look at whether some degree of ethical consideration might apply.

The fear of "runaway" xenobots also emerges from these discussions. Could a xenobot, if not carefully monitored, replicate in uncontrolled ways or interact with other organisms in unpredictable manners? The scientific community is already focused on building safety mechanisms into xenobot design, including limits on lifespan and reproduction, to prevent uncontrolled growth.

However, as their autonomy and complexity grow, the need for stringent ethical and safety oversight increases, underscoring the importance of a cautious approach to xenobot development.

Animal Rights and Environmental Ethics

Because xenobots are derived from frog cells, their creation touches upon ethical questions concerning animal rights.

Even though these cells do not come from sentient frogs and do not cause harm to adult animals, the process nonetheless involves the use of living biological materials.

Animal rights advocates may argue that any manipulation of animal cells for experimental purposes is ethically questionable, regardless of the sentience of the source material.

They may argue that creating xenobots for human use requires a careful ethical review to ensure respect for all forms of life, however simple or complex.

Environmental ethics come into play when considering the potential ecological impacts of xenobot deployment.

While xenobots could help clean up pollutants and provide other environmental benefits, releasing biological machines into natural ecosystems could disrupt delicate ecological balances in unforeseen ways.

Ethical guidelines, therefore, need to consider the possible environmental consequences, balancing potential ecological benefits against the risk of unintended harm to ecosystems.

The Risk of Dual-Use Technology

Xenobots also represent what is known as "dual-use" technology—technology that can be used for both beneficial and potentially harmful purposes.

While their primary applications are currently focused on health, environmental science, and research, xenobots could theoretically be repurposed for military or surveillance uses.

In a worst-case scenario, xenobots could be adapted for harmful purposes, such as biological espionage or biological warfare.

The dual-use nature of xenobots demands international ethical oversight, with regulations in place to prevent misuse and to ensure that their applications remain in alignment with ethical standards and public safety.

Regulatory and Ethical Frameworks for Future Xenobot Research

As xenobot technology continues to advance, there is a pressing need for robust regulatory and ethical frameworks to guide their development and application.

Currently, few laws exist that directly address the unique nature of xenobots, leaving scientists and ethicists in largely uncharted territory.

Regulatory bodies, bioethicists, and policymakers must work together to establish guidelines that cover both the potential benefits and risks of xenobot research.

Several organizations, including international bioethics councils and scientific regulatory agencies, are already beginning to explore potential guidelines for xenobot research.

Proposed frameworks emphasize transparency, public engagement, and careful risk assessment. Open dialogues between scientists and the public are essential to addressing ethical concerns and fostering trust in the responsible development of this technology.

The Path Forward: A Responsible and Ethical Future for Xenobots

Ultimately, xenobot research holds enormous promise but also requires a careful, measured approach. While the technology has the potential to deliver remarkable benefits, it also raises challenging questions about the nature of life, the limits of human intervention in biology, and the ethical responsibilities of science.

Ensuring that xenobot development aligns with ethical principles, respects environmental and animal welfare, and prioritizes public safety will be essential for navigating this uncharted ethical territory. Only through transparent, ethical practices can society harness the potential of xenobots while safeguarding against the risks, paving the way for a future where biological technology serves humanity in beneficial and responsible ways.

Delving Deeper into the Ethical Dilemmas Posed by Xenobots

Xenobots, the world's first living, programmable organisms, bring with them an extraordinary array of ethical questions and challenges. Made from frog cells and carefully arranged into tiny robotic shapes, these "living robots" can move, heal themselves, and even work together in certain scenarios. As these tiny, millimeter-sized life forms are designed to perform a range of tasks—from environmental cleanup to potentially medical applications—researchers, ethicists, and society at large find themselves at the crossroads of profound questions. What does it mean to create a life form designed for human purposes? Are we playing with forces beyond our understanding? Or, more pragmatically, how do we ensure that xenobots remain beneficial and ethical without crossing moral boundaries?

1. Defining Life and the Moral Status of Xenobots

One of the most pressing ethical dilemmas concerns the definition of life and whether xenobots should be granted a moral status similar to animals or even human cells. Although they are created from living frog cells, xenobots do not possess a brain, nervous system, or consciousness, nor do they resemble a traditional organism. They do not experience pleasure or pain and are highly limited in terms of sensory input.

Yet, even as non-sentient creations, xenobots challenge conventional ideas about life. Are they closer to robots or organisms? Should they have any rights or protections under animal welfare laws? While they are currently viewed as biological tools rather than beings, as researchers continue to refine them, xenobots may grow more complex. Some scientists worry that even a marginal increase in their capabilities could blur the line between mere tools and living entities deserving ethical consideration.

2. Consent and Manipulation of Living Cells

Another ethical question centers around the manipulation of living cells to create these artificial entities. Xenobots are crafted from frog embryos, which raises questions about the ethical sourcing and handling of biological material. While frog cells themselves do not raise the same moral concerns as human cells, manipulating any form of life introduces concerns about consent and the unintended consequences of tampering with natural processes.

By engineering life forms to fit specific roles or perform human-centered tasks, we're essentially manipulating the building blocks of biology to serve us. Some ethicists argue that this could create a dangerous precedent of engineering life without fully understanding or respecting the consequences. Unlike traditional robots made from silicon and metal, xenobots are alive, raising questions about whether it is ethical to design life forms purely for utilitarian purposes. Would it be ethical, for example, to create a version of a xenobot specifically tailored for hazardous waste disposal that might suffer as a result? Such questions remain open, underscoring the need for careful ethical consideration.

3. Potential for Uncontrolled Evolution

One of the biggest ethical fears surrounding xenobots is the potential for uncontrolled evolution. Although xenobots are currently limited in capability, some can self-replicate by clustering nearby cells into new copies. While this self-replication is simple and limited, it raises concerns about potential unintended mutations or adaptations. Could xenobots evolve beyond their intended functions? Could they develop harmful or invasive behaviors if released into the environment?

Scientists are working to minimize these risks, often designing xenobots to degrade over time or implementing mechanisms that prevent them from replicating outside controlled environments. However, even with such safeguards, the potential for accidental release and unintended consequences remains a topic of debate. Ensuring that xenobots do not harm ecosystems or become invasive species is essential, requiring strict containment protocols and oversight. Ethicists argue that researchers must develop a strong framework for assessing these risks to ensure that xenobots remain beneficial and benign.

4. Environmental Impact and Sustainability

While some propose xenobots for ecological cleanup, such as collecting microplastics from oceans or delivering targeted drugs to specific cells in the human body, their environmental impact could still be unpredictable. Xenobots are organic and biodegradable, but as they degrade, their components reenter the environment. How will their degradation products interact with existing ecosystems? Could they inadvertently introduce foreign biological material that disrupts delicate environmental balances?

Moreover, using living organisms to solve pollution problems raises a deeper ethical question: Are we using xenobots as a "quick fix" rather than addressing the root causes of environmental issues? Some environmental ethicists argue that relying on xenobots for ecological cleanup could distract from the need to reduce waste and consumption. While xenobots might help mitigate certain environmental challenges, we must consider their use carefully to avoid creating new problems in the process.

5. Implications for Human Health and Medical Applications

Xenobots have exciting potential for medical applications, such as delivering targeted therapies or repairing tissue damage. However, using living organisms within the human body introduces complex ethical challenges, particularly around safety, consent, and long-term effects. Xenobots could represent a new class of biomedical tools that blur the line between treatment and manipulation of the body.

When designing xenobots for medical applications, ethical concerns arise about patient safety and autonomy. How do we ensure that xenobots introduced into the body do not pose unintended health risks or lead to unforeseen complications? If xenobots are designed to alter or interact with human cells, could they cause unforeseen changes in the human body?

Moreover, as xenobot technology advances, patients and medical professionals must grapple with questions around informed consent and transparency. Ensuring that patients fully understand the risks associated with xenobot treatments is paramount.

6. The "Playing God" Debate and Human Responsibility

The concept of creating life for specific tasks has long been associated with concerns about humans "playing God." Xenobots push this debate further, as they represent a level of control over biological material that was once thought impossible. Critics argue that humanity's increasing power over life may lead to unintended moral and existential consequences.

From a practical standpoint, this debate speaks to the broader issue of human responsibility. Are we prepared to handle the ethical complexities that come with designing life? Can we predict the societal, environmental, and philosophical implications of deploying xenobots at scale?

While some argue that humans have always manipulated nature, others believe that xenobots represent a new level of intervention that could backfire if not approached with caution.

7. The Need for Ethical Oversight and Regulation

Given the ethical and environmental concerns surrounding xenobots, there is a pressing need for oversight and regulation. Some scientists advocate for a framework similar to those used for genetically modified organisms (GMOs) and synthetic biology.

A regulatory framework could provide guidelines on xenobot containment, research boundaries, and acceptable uses.

Establishing ethical oversight not only helps protect society from potential misuse of xenobots but also encourages responsible innovation.

Engaging with the public, policymakers, and ethicists can help ensure that xenobot research remains transparent and aligned with societal values.

By prioritizing ethical responsibility, the scientific community can harness the benefits of xenobot technology while addressing its challenges thoughtfully.

Striking a Balance

As the potential applications for xenobots expand, the ethical dilemmas they pose will only grow more complex. Xenobots offer a vision of the future where living organisms are crafted to perform tasks and solve problems, but they also challenge our understanding of life, responsibility, and the natural world.

Embracing the potential of xenobots requires careful thought, not only about their scientific possibilities but also their ethical implications.

In navigating these questions, we find ourselves on the threshold of a new era in biology—one that demands wisdom, restraint, and a profound respect for the life we are now capable of creating.

Navigating Complex Moral Questions Surrounding Living Machines

In recent years, scientific research has entered a new frontier by creating "living machines" – tiny, biologically engineered entities known as xenobots. These remarkable organisms, built using stem cells from the African clawed frog (Xenopus laevis), represent the fusion of biology and technology.

Xenobots possess unique capabilities, from movement and self-repair to limited problem-solving. This blend of cellular biology and artificial intelligence has sparked both excitement and deep ethical questions.

The Birth of a New Lifeform: Defining Living Machines

Xenobots blur the boundary between life and machine. They are neither entirely biological nor mechanical, but exist in a "gray zone" between these categories. Developed by scientists from Tufts University and the University of Vermont, xenobots consist of living cells that are assembled into programmable entities, capable of moving autonomously, carrying small payloads, and even organizing themselves for specific tasks.

Unlike conventional robots made of metal or plastic, xenobots are biodegradable, and since they're composed entirely of living tissue, they can repair minor damage by regenerating their cells.

This breakthrough raises a fundamental question: what does it mean to be alive? While xenobots are technically living (they are composed of live cells and capable of limited self-repair), they do not possess nervous systems, consciousness, or emotions. Yet, the fact that they are living tissue with a degree of autonomy introduces complex ethical considerations. Are xenobots a form of life deserving moral consideration? Or are they simply tools to be used for human purposes?

Potential Benefits: Medicine, Environmental Impact, and Scientific Exploration

Before delving into the moral quandaries, it is important to recognize the immense potential benefits xenobots could offer. In medicine, for example, xenobots could be deployed to deliver drugs precisely where they're needed, reducing side effects and improving patient outcomes. They might also aid in removing blockages within blood vessels or clearing up toxic substances within the body, performing functions that traditional medicine struggles with.

In environmental terms, xenobots could become bioengineered agents that clean up plastic pollution or tackle oil spills in ways that mechanical robots cannot. Since they are biodegradable, they offer a more sustainable alternative to plastic or metal-based machines, reducing environmental footprints and promoting eco-friendly solutions.

Moreover, studying xenobots enables scientists to better understand cellular behavior, regeneration, and even the mechanisms of life itself.

These potential applications highlight how xenobots could benefit society, but they also give rise to concerns about the ethical limits of their use.

Should xenobots be allowed to operate autonomously within the human body or natural ecosystems? And if so, under what conditions?

As with any powerful new technology, the potential for misuse is real and must be addressed.

The Ethics of Creation: Playing with Life?

The process of creating xenobots raises a question as old as humanity: are we "playing God" by building new forms of life? Historically, innovations like genetic engineering and cloning have prompted similar concerns.

With xenobots, scientists are not merely editing DNA but constructing entirely new biological entities with no equivalent in nature.

This intentional act of creating life – albeit a lifeform without consciousness – challenges traditional understandings of life and raises moral and spiritual questions.

Some argue that creating xenobots is ethically permissible because they lack self-awareness and cannot experience pain or suffering. Others worry that we are setting a dangerous precedent by creating organisms designed to follow commands.

What if future iterations of xenobots develop beyond current limitations?

As our understanding of cellular programming improves, the possibility of designing more complex living machines becomes very real, forcing society to confront the uncomfortable question: at what point does a xenobot become "alive" in a moral sense?

Autonomy and Control: The Risk of Unintended Consequences

A critical ethical issue lies in the autonomy of xenobots. Although current xenobots are programmed to perform specific tasks and have limited lifespans, the line between controlled programming and autonomy could become increasingly blurred as the technology advances. Autonomous xenobots might be able to adapt to their environments, reproduce, or perform tasks without direct human supervision. This independence could lead to unintended consequences, particularly if xenobots operate in complex ecosystems where predicting their effects is challenging.

For example, if xenobots were introduced into a marine environment to break down microplastics, what safeguards would prevent them from affecting other organisms? How can we ensure that they don't disrupt existing ecosystems in unexpected ways?

These questions highlight the need for strict oversight, ethical guidelines, and clear limitations on the deployment of xenobots to minimize risks.

Another key consideration is the potential misuse of xenobot technology. In the wrong hands, xenobots could be engineered for harmful purposes, such as espionage, bio-warfare, or unethical surveillance.

This possibility demands that regulations and safety protocols be established before xenobot technology becomes widespread. Transparency, accountability, and ethical governance are essential to ensuring that xenobots are used responsibly.

Moral Status and Rights: Should Xenobots Have Protections?

One of the most provocative questions surrounding xenobots is whether they deserve moral consideration or rights. While current xenobots lack the cognitive and sensory capacities that would typically warrant moral status, the potential for more advanced living machines raises questions about future moral obligations.

For example, if xenobots could eventually feel sensations or develop rudimentary forms of learning, would they merit ethical protections?

Some ethicists argue that, at a minimum, xenobots should be treated with respect, particularly when they involve the use of animal-derived cells. This respect might mean limiting experiments that could cause harm, even if xenobots are not conscious. Others believe that granting moral status to xenobots could create a slippery slope, leading to demands for rights for increasingly complex synthetic lifeforms.

This debate touches on profound issues of empathy, responsibility, and the expanding definition of sentience.

Charting a Path Forward: Ethical Guidelines for Living Machines

As xenobot technology progresses, society must establish ethical guidelines that account for both the unique nature of these organisms and the potential impacts they may have. Guidelines should address the boundaries of xenobot capabilities, acceptable applications, and ethical practices for research and deployment. Regulators, scientists, and ethicists must collaborate to create frameworks that balance innovation with caution, emphasizing transparency and accountability.

One approach could involve setting up an ethical review board specific to xenobot research, ensuring that experiments align with social and environmental values. Public engagement and education are also essential to fostering a well-informed dialogue about the future of xenobot technology. By involving diverse perspectives – from scientists to philosophers to the general public – society can build a more nuanced understanding of living machines and the ethical principles that should guide their development.

In conclusion, xenobots and similar living machines represent a transformative scientific breakthrough with the potential to reshape medicine, environmental science, and our understanding of life itself. Yet, this innovation also introduces complex moral questions that challenge long-held beliefs about life, autonomy, and responsibility. As we continue to explore this exciting frontier, society must remain vigilant, embracing both the promise and the ethical challenges of living machines with thoughtfulness and care. The path forward will require a delicate balance between the pursuit of knowledge and the respect for life in all its evolving forms.

Chapter 29: Xenobots and the Quest for Artificial Life

Imagine a tiny, living robot, one so small it could navigate through the human body, swim through veins, and potentially deliver medicine to specific cells. Picture it searching for disease, cleaning up pollutants, or even repairing tissue. This isn't science fiction—these are xenobots, the world's first living robots, created from biological tissue in a lab. Named after the African clawed frog (Xenopus laevis), from which their cells are derived, xenobots represent an exciting leap in the search for artificial life.

Xenobots are made from living cells rather than traditional synthetic materials. The creation of these tiny, programmable biological entities is the result of a collaboration between computer scientists and biologists. In 2020, researchers at the University of Vermont and Tufts University used cells from frog embryos to create these programmable organisms. Unlike conventional robots made of metal or plastic, xenobots are entirely biological, which gives them unique properties and potential advantages for environmental and medical applications.

The creation of xenobots involves a fascinating process. Researchers start with frog embryos and extract two specific types of cells: skin cells and heart muscle cells. Skin cells provide a sturdy structure for the xenobot, while heart muscle cells act as tiny motors, contracting and expanding to produce movement. By assembling these cells in specific configurations, scientists can create tiny, programmable organisms that exhibit behaviors based on their cellular arrangement.

The development process also involves the use of powerful computer algorithms. Researchers employ an AI-based simulation to design the optimal shape and structure of the xenobot. By feeding the AI thousands of possible cell configurations, it learns which designs are most effective for specific tasks. These designs are then translated into real biological forms, with the final product being a living, moving organism less than a millimeter in size.

What makes xenobots truly fascinating is that they are programmable. This doesn't mean they can run software code like a traditional computer, but their design and structure can dictate their behaviors.

For instance, xenobots can be created to move in a straight line, circle, or specific patterns based on how their heart muscle cells are arranged. They can also be designed to work collectively, with multiple xenobots coordinating their movements, much like a swarm of ants or bees working together.

One of the significant achievements in xenobot development is their ability to self-heal. When damaged, these living machines can regenerate their tissues. This self-repair capability is a natural characteristic of the biological cells they are made from, and it could have profound implications for the durability and longevity of future artificial lifeforms.

Unlike traditional machines that wear out or break over time, xenobots can theoretically last longer with minimal maintenance because of this regenerative property.

Another fascinating feature of xenobots is that they are biodegradable. Made entirely of organic materials, they naturally decompose over time, posing no risk of long-term pollution.

This could make xenobots an ideal solution for tasks like cleaning up microplastics in the ocean or removing harmful pollutants from contaminated environments, as they wouldn't leave any toxic residue behind.

Xenobots offer the potential to revolutionize medicine. Since they are small enough to travel through bodily fluids, they could one day deliver targeted drugs directly to cancer cells, avoiding the need for harmful chemotherapy that affects the entire body.

Researchers are also exploring the possibility of using xenobots to clear arterial plaque, heal wounds, and even conduct microsurgeries that are currently impossible for human surgeons. The prospect of tiny, biologically safe robots traveling through our bodies, healing as they go, is a powerful vision of future healthcare.

However, as exciting as xenobots are, they also raise ethical and philosophical questions. Since xenobots are alive but programmable, they challenge our definition of life itself. Are they organisms in the traditional sense, or are they machines? Can they be considered a new form of life, or are they merely tools for human use?

These questions are not only scientific but also ethical, as they touch upon our understanding of what it means to be alive and the responsibilities we bear in creating new forms of life.

Critics also worry about potential misuse or unintended consequences. For example, what if xenobots evolved beyond our control? While xenobots today are simple and have limited lifespans, future versions might become more complex, leading to scenarios where they could behave unpredictably.

Moreover, some fear that xenobots, if engineered for malicious purposes, could be weaponized, although scientists are currently working within strict ethical guidelines to prevent such outcomes.

To address these concerns, the scientific community is implementing strict protocols and oversight. Each xenobot is designed with a built-in lifespan and is incapable of reproducing, ensuring that it cannot proliferate uncontrollably.

Furthermore, researchers are investigating ways to control and monitor xenobot behavior, ensuring that any future developments align with societal and ethical standards.

Beyond practical applications, xenobots also help us understand the fundamental principles of life. By creating programmable organisms, scientists are uncovering clues about the origins of multicellular life.

Understanding how cells can be organized to perform specific functions sheds light on the evolutionary processes that shaped life on Earth. In essence, xenobots serve as simplified models for studying how different cells cooperate, communicate, and develop into complex organisms.

The quest for artificial life, as exemplified by xenobots, pushes the boundaries of biology, robotics, and artificial intelligence. It represents a convergence of scientific disciplines working together to unlock new possibilities.

As we continue to refine the technology, xenobots may pave the way for more advanced biohybrid systems—organisms that combine biological tissues with synthetic materials to create even more sophisticated and adaptable lifeforms.

Ultimately, the development of xenobots is just one step in a broader journey toward understanding and creating artificial life. While they are far from the sentient, autonomous robots depicted in popular culture, xenobots offer a glimpse into a future where living machines could transform medicine, environmental conservation, and perhaps even our understanding of life itself.

The journey of xenobots is still in its early stages, but the possibilities are vast. These tiny living robots could usher in a new era of biotechnology, where artificial life is not just a concept, but a tool for improving our world.

As research progresses, we may find ourselves on the brink of a revolution that redefines life itself—transforming xenobots from a laboratory experiment into a cornerstone of 21st-century science and technology.

Philosophical Reflections on Xenobots and Artificial Life

Imagine a world where life can be designed, assembled, and even reconfigured—this is the reality that xenobots, or "programmable living organisms," are beginning to unlock. Created from the stem cells of frogs (specifically, the African clawed frog, Xenopus laevis), xenobots are a groundbreaking new form of life that blur the boundary between the natural and artificial.

They are not mechanical robots nor traditional living organisms, but rather a synthesis of the two: tiny biological machines made up of living cells that can be "programmed" to perform specific tasks.

The development of xenobots has led to a new frontier in science and technology, as well as a space rich for philosophical inquiry. At its core, the creation of xenobots challenges our notions of life, intelligence, and the ethical responsibilities that arise with artificial life forms.

By examining xenobots, we encounter new questions: What does it mean to be alive? How do we define purpose and agency in life that has been engineered rather than evolved? And most importantly, what responsibilities do we have as creators of new forms of life?

Redefining Life

One of the fundamental philosophical questions raised by xenobots is the nature of life itself. Traditionally, life has been associated with characteristics like metabolism, growth, reproduction, and response to stimuli. Xenobots fit some of these criteria, as they can move, respond to their environment, and even repair themselves.

However, they are not products of natural evolution. Instead, they are created by humans through the manipulation of cellular behavior, using advanced algorithms and biological engineering. This brings into question whether xenobots should be categorized as truly "alive" or as an advanced, bio-hybrid form of machinery.

The question is complex. If life can be defined by the presence of certain characteristics or behaviors, then xenobots might indeed qualify. Yet, life as we understand it traditionally involves a genetic lineage that has undergone evolutionary pressures over millions of years. Xenobots do not share this lineage; their "purpose" has been pre-programmed by humans.

This artificial purpose contrasts with the evolutionary "purpose" seen in natural organisms, whose behaviors and structures evolved to enhance survival and reproduction.

Philosophically, this invites us to consider if xenobots, devoid of natural evolutionary history, are truly alive or if they are simply sophisticated biological machines.

Agency and Autonomy

Xenobots raise questions about agency and autonomy, especially regarding the concept of "programming." When we create a xenobot, we essentially set its parameters and purpose. For instance, xenobots have been designed to move in specific patterns, clear away microscopic particles, or even work in groups. These tasks are not chosen by the xenobots but are instead embedded into their very design. Does this lack of autonomous choice negate the concept of agency? Can something with no intrinsic purpose other than what has been imposed by its creator possess any form of autonomy?

In some ways, this question parallels philosophical debates about free will in humans. If our behaviors can be determined by our biological and environmental influences, does this mean that we, too, lack autonomy? Xenobots offer a mirror to human agency in this regard. They demonstrate how specific behavioral patterns can be instilled into life forms through design.

This raises an interesting perspective: autonomy may not be a binary state but a continuum. Xenobots, though "programmed," can respond to certain environmental changes, adjust their movements, and heal themselves. They show that agency might not require complete freedom from influence but may instead be the ability to react and adapt within certain constraints.

Ethical Considerations

The creation of artificial life forms, such as xenobots, raises pressing ethical questions. Is it morally acceptable to create life with predetermined purposes? Do we bear responsibility for the actions and consequences of these artificial life forms? Xenobots are currently limited in their functions and capabilities, but as technology advances, it is likely that more sophisticated versions could emerge.

The potential for xenobots to be used in various applications, from medicine to environmental cleanup, introduces a need to establish ethical guidelines on their use and limitations.

One significant ethical concern is the potential suffering of artificial life. While xenobots currently lack a nervous system, it is not implausible that future bio-hybrids might have some form of neural network. Would they then have the capacity to experience sensations, including pain? If so, we would need to address whether it is ethical to use them in ways that might cause suffering, even if they are "designed" for a particular purpose. The line between organic life and artificial creation is blurring, and with it, the ethical considerations that govern our treatment of living beings may need re-evaluation.

Moreover, xenobots highlight the issue of environmental ethics. If we deploy these bio-hybrid organisms into natural ecosystems, we must carefully consider the potential ecological impacts. Even if xenobots are programmed to disassemble after a certain period, there is no guarantee that their behavior won't have unintended consequences on local ecosystems. As creators, humans would bear moral responsibility for any harm caused by these artificial life forms. This responsibility extends not only to individual xenobots but to the larger ecological systems they might impact.

Xenobots and Human Identity

The creation of xenobots also leads us to reflect on our own identity. Throughout history, humans have distinguished themselves as beings capable of self-reflection, creativity, and complex moral reasoning. By creating xenobots, we are asserting our ability to create life, albeit in a limited way. This act can be seen as an expression of our creativity and scientific ingenuity, but it also raises questions about what it means to be human.

If we can create life that mimics certain aspects of human behavior and intelligence, where does that leave us in the hierarchy of existence? Xenobots prompt us to reconsider the uniqueness of human life. They remind us that intelligence, adaptability, and the capacity to perform complex tasks may not be as exclusive to humans as once thought. This new perspective invites humility and a recognition that life, in its various forms, may possess many of the traits we once thought unique to ourselves.

A New Frontier of Responsibility

The philosophical implications of xenobots extend beyond questions of life and intelligence. As we develop more sophisticated bio-hybrid technologies, we enter a realm of increased responsibility. We are not merely observers of the natural world; we are now shapers of it. The power to create life—or at least a version of it—brings with it an obligation to consider the ethical, environmental, and existential consequences of our actions.

Ultimately, xenobots are more than just a scientific achievement; they are a symbol of humanity's growing power to intervene in the fabric of life itself. They invite us to reflect on the kind of future we want to create, the boundaries we are willing to cross, and the responsibility we bear toward the world and the life forms within it—natural or artificial. As we continue to explore this frontier, our understanding of life, agency, and ethics will inevitably evolve, and with it, our understanding of what it means to be human.

Challenging Our Definitions of Life and Existence

Xenobots, the tiny, programmable biological machines crafted from frog cells, have been stirring scientific circles and popular imaginations alike. Developed by researchers using African clawed frog (Xenopus laevis) cells, these living organisms are neither strictly robots nor traditional biological organisms, and this ambiguity places them at the edge of our understanding of life itself. By examining their creation, capabilities, and implications, xenobots challenge fundamental concepts about what it means to be alive, to evolve, and to possess an independent existence.

Xenobots are microscopic structures assembled from living cells, particularly skin and cardiac (heart) cells. The heart cells give xenobots the ability to move, while the skin cells provide structure. Scientists designed them with algorithms to create cell formations that can perform simple tasks like moving, carrying small objects, and even self-healing. Unlike most robots, xenobots aren't metal-based or machine-like in appearance. They are living, flexible, and biodegradable. But they also don't fit into our classic idea of a living organism, which makes them an intriguing point of study for understanding life itself.

Life at the Edge of Biology and Engineering

The very existence of xenobots puts them at the intersection of biology, engineering, and computer science. In essence, xenobots are "programmed" not through microchips or software but through biological manipulation. By altering their cellular structure, scientists can direct them to perform specific tasks, which allows xenobots to exhibit purposeful behavior. This ability to modify and reconfigure cells to perform tasks goes against our traditional understanding of how organisms work and develop in nature. While natural organisms evolve through reproduction and natural selection, xenobots are created and modified in the lab, with scientists deciding their design and purpose.

This artificial manipulation of cellular life raises profound questions: If an organism does not evolve in nature, can we still call it "alive"? If its purpose and structure are entirely determined by humans, does it have its own "existence," or is it merely a tool? Xenobots occupy a unique space that blends organismal life with machine functionality, challenging definitions of both.

Defining "Life" in the Age of Xenobots

The scientific community has long debated a working definition of life. Traditionally, life is considered to have certain characteristics, such as metabolism, reproduction, growth, adaptation, and response to stimuli. Xenobots, however, do not entirely fit within these parameters. They are made from living cells, so they can respond to their environment and, in some cases, self-heal. However, they do not have metabolism in the traditional sense, nor do they reproduce or evolve over generations. Instead, scientists directly control their design and functions.

Because xenobots blur these criteria, they prompt scientists to question the usefulness of current definitions of life. Are xenobots "alive" because they are made of living cells, or are they simply tools controlled by an external force? If an entity can exist, perform tasks, and interact with its environment without having a metabolism or the ability to reproduce, is it not still in some way "alive"? By presenting these gray areas, xenobots encourage us to explore more flexible, inclusive definitions of life that might someday accommodate synthetic biology and artificial life forms.

The Question of Existence and Autonomy

One of the most profound challenges xenobots pose is to our concept of existence. Traditional life forms operate with a degree of autonomy: they grow, seek nutrients, and reproduce independently.

Even the simplest organisms, like bacteria, have life cycles and processes that allow them to perpetuate themselves.

Xenobots, in contrast, rely entirely on scientists to exist. Their "life" begins in a lab, and they lack the internal systems to sustain or propagate themselves independently.

This dependency raises philosophical questions about the nature of existence and autonomy. Xenobots do not have DNA-based evolutionary histories or biological imperatives that direct them to seek out resources, reproduce, or adapt.

They challenge the assumption that life is defined by a need to perpetuate itself, suggesting that existence might also encompass entities that do not follow these evolutionary imperatives.

However, this lack of autonomy also highlights a potential new category of life: something that is "alive" but not "self-directed." This opens up a discussion about whether we might someday consider other artificial life forms to have "existence," even if they lack traditional autonomy.

Ethics, Identity, and the Boundaries of Being

The creation of xenobots raises ethical questions about identity, purpose, and our responsibility as creators. Are xenobots a new form of being, or are they simply extensions of human will? If they can be programmed to perform tasks, who holds the ethical responsibility for their actions and consequences?

Moreover, xenobots make us question whether we have the right to define what should or should not be considered "alive" based on current biological standards.

Xenobots also invite comparisons with artificial intelligence, where similar debates exist over machine autonomy, consciousness, and ethical design.

However, xenobots go beyond AI by embodying living cells. They provoke us to consider whether manipulating living cells to create new forms challenges moral boundaries.

If xenobots eventually evolve more complex functionalities, we may be compelled to address whether they deserve consideration as a new form of life with inherent value.

Potential Implications for Future Life Forms

The study of xenobots could lay the foundation for developing more advanced forms of artificial life, with potential applications in medicine, environmental management, and beyond.

Xenobots could be designed to deliver drugs to specific parts of the human body, clean up environmental waste, or perform other complex tasks that require a combination of mobility and adaptability.

In the future, we might see xenobots with greater autonomy or the capacity to make independent decisions, moving us closer to the possibility of synthetic, artificially designed organisms.

The implications of xenobot research extend beyond practical applications.

The success of xenobots suggests that humans can, in fact, design and shape biological entities in ways previously limited to natural evolution.

This shift could fundamentally alter our view of evolution, genetics, and the potential for life beyond Earth.

If life can be "engineered" rather than simply arising through natural processes, the possibility of creating synthetic life forms expands.

A Redefinition of Life Itself?

Xenobots embody a new frontier that challenges long-held assumptions about life, existence, and what it means to be alive.

They invite us to reconsider how we define and interact with life forms and push the boundaries of what it means to create and control living matter.

As we delve further into synthetic biology, xenobots remind us that the line between biology and technology is becoming ever more porous.

In the coming decades, the study of xenobots may offer new insights not only into life on Earth but also into how we might understand or even create life beyond it.

By pushing us to reexamine our definitions and assumptions, xenobots pave the way for an era where life, in all its forms, might be understood as something we can design, shape, and perhaps even evolve by choice rather than by chance.

Chapter 30: The Ever-Evolving Xenobot Revolution

Xenobots, often described as "living robots," have captured imaginations and scientific intrigue since their inception. These tiny, programmable biological robots are crafted from the cells of the African clawed frog (Xenopus laevis)—thus their name, "xenobot." Their creation and development mark a fascinating leap in bioengineering, one that blurs the boundaries between biology, robotics, and artificial intelligence. The field of xenobot research is a rapidly evolving frontier that could profoundly alter medicine, environmental science, and even ethics.

Origins of Xenobots

The concept of xenobots was born in 2020, through a collaborative effort between scientists at Tufts University and the University of Vermont. This pioneering team was led by computer scientist and roboticist Joshua Bongard and biologist Michael Levin, who explored how living cells could be repurposed to perform programmable actions. By using stem cells from Xenopus laevis, the team developed biological entities that could be shaped and instructed to perform simple tasks.

This breakthrough was made possible by advances in computer algorithms, which allowed researchers to simulate countless potential forms for these organisms, testing them digitally before crafting physical models.

Xenobots are created by gathering frog skin and heart cells, which are then sculpted into specific configurations. Skin cells provide structural support, while heart cells, which contract rhythmically, generate movement.

This unique assembly means xenobots can move, respond to environmental stimuli, and potentially carry tiny payloads—all without a traditional nervous system or DNA-driven purpose.

How Xenobots Function

The beauty of xenobots lies in their simplicity. Unlike mechanical robots, xenobots are living entities that self-organize based on their cellular composition. The heart cells' contractions act as tiny motors, enabling the xenobots to propel themselves in water. Researchers leverage artificial intelligence to predict and optimize xenobot forms, simulating various configurations in virtual models. This process allows scientists to fine-tune xenobot shapes for specific functions, whether it be movement, object transport, or self-healing capabilities.

Xenobots operate in water and are designed to work as individual units or in collective formations. In initial experiments, these biological robots demonstrated basic swarm behaviors, such as gathering particles or working together to move larger objects. Importantly, xenobots are biodegradable, as they naturally break down after a certain period, reducing environmental impact—a key advantage over traditional robotic devices.

Medical and Environmental Potential

One of the most promising applications for xenobots lies in the medical field. Their ability to move in targeted ways opens the door for novel drug delivery methods within the human body. Imagine a xenobot designed to carry a medication to a specific location, such as a tumor or an infection site, where it could deliver treatment precisely, reducing the side effects of traditional drug delivery methods. Xenobots might one day assist with wound healing or even be used to remove harmful substances from the body.

In the environmental realm, xenobots could serve as effective tools for pollution cleanup. For example, they could be designed to capture microplastics in water sources, a pervasive environmental problem that traditional methods have difficulty addressing. Since xenobots are organic and biodegradable, they provide an eco-friendly alternative to plastic-based cleanup mechanisms. Researchers have proposed that fleets of xenobots could patrol lakes, rivers, or oceans to collect pollutants before degrading naturally, leaving no toxic residue.

The Self-Replication Phenomenon

In recent developments, xenobots have demonstrated an unexpected ability: self-replication. When placed in specific conditions, xenobots can gather loose cells to assemble new versions of themselves. This form of replication differs from typical biological reproduction and is closer to assembly, as if a robot were building a copy of itself from scattered parts. While this capability is rudimentary, it underscores the adaptability of living cells and the potential for creating self-sustaining biological systems. Self-replication opens up exciting possibilities for creating autonomous fleets of xenobots capable of maintaining or expanding their presence in a given environment. However, it also raises questions about containment, control, and ethical boundaries in bioengineering.

Ethical and Regulatory Challenges

The rapid progress of xenobot technology raises ethical questions that society must confront. Since xenobots are living organisms created through human intervention, concerns arise around their rights and welfare. Should xenobots be treated as tools, or do they deserve a new category of ethical consideration as life forms engineered for human purposes?

Another major concern is containment and control. If xenobots are released into the environment or introduced into the human body, mechanisms must be in place to prevent unintended consequences. Autonomous xenobot swarms could, theoretically, cause harm if they behave unpredictably. Developing regulations that ensure their safe usage while fostering innovation is crucial as xenobot technology evolves.

Future Horizons of Xenobot Research

The next stages of xenobot research involve exploring more complex tasks and expanding their capabilities. For instance, scientists are experimenting with xenobots that can respond to chemical signals, potentially enabling them to detect environmental pollutants or sense diseased tissues in the body. Research is also focusing on improving xenobot durability, allowing them to operate in diverse conditions, including extreme temperatures or high salinity.

One area under consideration is incorporating neural-like properties into xenobots, potentially creating living robots capable of more advanced forms of sensing and decision-making.

Though it's a distant possibility, these neural xenobots could someday learn to respond to various stimuli or adapt to new environments, making them ideal candidates for missions that require autonomous adaptation, such as space exploration or disaster relief operations.

The Broader Impact of Xenobots

The xenobot revolution has implications that stretch beyond immediate applications. The methods used to design and deploy xenobots could influence other fields of bioengineering, leading to innovative ways to harness and control biological cells.

Concepts like programmable cells, tissue engineering, and synthetic biology are likely to advance in parallel, offering more refined methods to build biological systems with desired functions.

Moreover, the interdisciplinary nature of xenobot research—bringing together biology, robotics, AI, and ethics—may inspire similar cross-disciplinary ventures in other areas of science. Such collaborations are essential for tackling complex global challenges, from environmental degradation to healthcare.

Embracing the Future of Bioengineering

The xenobot revolution is a fascinating chapter in the ongoing story of scientific innovation. By creating programmable, biodegradable biological robots, scientists have laid the groundwork for a future where living machines work alongside us, tackling challenges in medicine, ecology, and beyond. As xenobots evolve, so too must our approach to ethics, regulation, and interdisciplinary research, ensuring that we harness these tiny biological marvels responsibly.

In time, xenobots may become part of our everyday lives, invisible allies helping to heal bodies, restore environments, and perhaps redefine what it means to build and create life.

The Journey from Inception to the Future

The world of artificial intelligence, robotics, and biotechnology has experienced numerous revolutionary breakthroughs over the years, but few have sparked as much fascination as xenobots. These tiny living robots—created from the cells of the African clawed frog, Xenopus laevis—represent a unique blend of biological and synthetic systems.

Their development is not only a marvel of scientific ingenuity but also a potential doorway to a future where technology and biology work together in unprecedented ways.

Origins and Conceptual Foundations

The journey of xenobots began with a simple yet ambitious question: Could biological tissues be used to create programmable living machines? In 2020, researchers at Tufts University and the University of Vermont achieved an extraordinary breakthrough.

By manipulating embryonic frog cells, they engineered a new kind of lifeform that could move, perform tasks, and even self-repair. Named "xenobots" after the Xenopus frog, these creations pushed the boundaries of biology and robotics.

At the heart of the xenobot project was a mix of AI, robotics, and developmental biology. A critical element in the design was the use of an evolutionary algorithm, which ran simulations to explore millions of possible configurations.

The AI explored thousands of potential designs, eventually identifying ones that could best perform specific tasks—such as moving in a particular direction, gathering particles, or even working cooperatively.

Scientists then recreated the optimal designs by assembling frog skin and heart cells into tiny living robots.

These heart cells provided the contractions necessary for movement, while the skin cells gave structural support.

How Xenobots Are Built and Function

Creating a xenobot involves isolating and manipulating stem cells from frog embryos. Under specific conditions, these cells self-organize, forming structures that resemble simple organisms capable of movement. The building blocks, heart and skin cells, work synergistically: heart cells provide the mechanical motion, and skin cells form the protective scaffold. This combination gives xenobots a unique, flexible form of movement that can be adjusted based on cellular configuration and structure.

Once created, xenobots do not rely on food, fuel, or an external power source. They function autonomously, utilizing stored energy within their cells. After roughly a week, the energy reserves deplete, and xenobots disassemble and degrade naturally, leaving no waste or toxic residue. This biodegradable nature is particularly significant for applications in medicine and environmental cleanup, as they would not contribute to pollution or long-lasting waste.

Initial Applications and Milestones

Early experiments with xenobots demonstrated a few of their remarkable capabilities. They were able to navigate through water, push small particles into piles, and even demonstrate a form of "collective behavior" when grouped. These initial trials showcased the potential of xenobots as micro-cleaners that could gather microplastics in oceans, clean up toxic waste, or even deliver drugs to targeted areas within the human body.

As xenobot designs evolved, so did their abilities. In 2021, scientists observed an even more groundbreaking phenomenon: xenobots that could "self-replicate" under certain conditions. These tiny organisms displayed a form of biological "kinesis" that allowed them to create new generations of xenobots by scooping up loose cells and forming them into new structures.

This ability to self-replicate, albeit in a controlled manner, opened new questions in cellular self-organization, pushing scientists to explore whether xenobots could develop more complex forms of autonomous reproduction.

Challenges and Ethical Considerations

Despite their promise, xenobots raise several ethical and safety concerns. One major challenge is predictability: as with any biological system, there is an inherent unpredictability in how xenobots will behave in different environments, especially over extended periods.

While current designs degrade naturally, future iterations may require tighter control to prevent unintended consequences.

Ethical concerns also surround the creation of lifeforms that do not naturally exist in the world. Though xenobots are not sentient, questions arise about the extent to which humans should manipulate living cells and direct their behavior.

Additionally, concerns about the possible misuse of xenobots—for instance, in unauthorized surveillance or biological experimentation—underscore the need for rigorous oversight and ethical guidelines.

To address these concerns, researchers are developing frameworks to ensure that xenobot research aligns with principles of transparency, responsibility, and public welfare.

Regulatory frameworks, ethical guidelines, and risk assessments are necessary to prevent misuse while allowing society to reap the potential benefits of this emerging technology.

The Expanding Horizons of Xenobot Technology

The future of xenobots is full of promise. Current research is exploring ways to make them more complex and multifunctional.

For example, by incorporating sensory cells, researchers hope to create xenobots capable of responding to environmental signals like temperature, chemical gradients, or light.

This would make them more versatile and enable applications in biosensing and environmental monitoring.

Medical applications also hold great potential. Xenobots could be designed to deliver drugs directly to specific cells or tissues, making treatments more effective and reducing side effects. They could also be engineered to seek out and destroy cancer cells, removing harmful pathogens or repairing tissues.

The field of regenerative medicine, in particular, stands to benefit immensely as xenobots might one day be able to replace damaged cells, assisting with healing and tissue repair.

As xenobot research progresses, another exciting avenue is biohybridization—integrating xenobots with other types of cells or synthetic components. Hybrid xenobots could combine the best features of biological and synthetic systems, leading to even more versatile, robust, and controllable living machines.

For instance, combining xenobot designs with neural cells might enable the development of biohybrid devices that can sense, process, and react to stimuli in sophisticated ways.

Looking Ahead: The Future of Xenobots and Society

While still in their early stages, xenobots symbolize a growing field where biological and synthetic systems converge. Future research could extend their lifespans, incorporate complex behaviors, and broaden their application scope.

Xenobots are also a gateway for furthering our understanding of cellular self-organization, tissue engineering, and bio-computation.

As researchers continue to expand xenobot capabilities, society must decide how to balance innovation with caution.

Xenobots have the potential to improve environmental sustainability, health, and technology in transformative ways, but only if managed responsibly.

The future of xenobots will be a collaboration between scientists, ethicists, policymakers, and the public.

In summary, xenobots mark a fascinating step forward in merging biology with technology. From cleaning polluted waters to potentially treating diseases, they demonstrate how biology's fundamental building blocks can be harnessed for human benefit.

As xenobots evolve, so too will our understanding of what it means to create programmable life, inviting humanity to contemplate the role of living machines in a rapidly changing world.

The Evolution and Impact of Living Machines

Imagine tiny, programmable organisms that blur the line between machine and living being, capable of self-repair, independent motion, and environmental adaptation.

This is the world of xenobots—living, self-assembled biological robots that carry vast potential to reshape science, medicine, and our understanding of life itself.

As we speculate on the continued impact and evolution of these living machines, a fascinating array of possibilities and questions emerges. From medical marvels to ecological applications, xenobots are opening doors that will change the future of technology, biology, and perhaps even humanity.

The Birth of Xenobots

Xenobots were first developed from the embryonic cells of the African clawed frog, Xenopus laevis. Unlike traditional machines made from metal or plastic, xenobots are entirely biological, assembled from organic cells that communicate, repair themselves, and have a limited lifespan, reducing the risk of environmental waste.

Researchers designed xenobots by using computational algorithms to determine optimal shapes and behaviors based on specific goals, such as movement or the transport of microscopic cargo.

This pioneering work demonstrated how a combination of biology, robotics, and artificial intelligence could create programmable organisms, a concept that was once purely science fiction.

Exploring Xenobot Capabilities and Applications

Since their debut, xenobots have shown promise in various fields. In medicine, for instance, they could be used as delivery systems to transport drugs precisely to diseased or damaged areas in the body. Their ability to move and operate at the cellular level gives them a unique advantage in areas where conventional medical tools cannot reach, such as delivering chemotherapy drugs directly to tumors or clearing clogged arteries.

Beyond medicine, xenobots hold potential for environmental applications. They could be designed to detect and remove toxins from contaminated water sources, breaking down harmful substances on a molecular level. Since xenobots are biodegradable, they could perform these tasks without leaving behind harmful residues.

Additionally, they might be adapted to help clean up microplastics from oceans or aid in the biodegradation of pollutants, creating a natural, effective cleanup crew.

In research settings, xenobots offer a new platform for studying fundamental biological processes. By observing how they self-assemble, heal, and perform tasks, scientists can gain deeper insights into cell biology and tissue engineering.

These insights may lead to advances in regenerative medicine, where growing complex tissues or organs from scratch becomes feasible.

Speculating on Future Evolutions of Xenobots

The xenobot technology we see today is only the beginning. As this field progresses, several avenues for the evolution of xenobots are plausible. For instance, future xenobots could be programmed for even more complex behaviors by incorporating neural-like cells, giving them decision-making capabilities that resemble simple brains. Imagine a xenobot that can assess its surroundings, determine the most effective path to a target, and adapt to obstacles or changing environments. This would enable xenobots to perform more sophisticated tasks and even work autonomously within complex biological or environmental systems.

Another potential evolution involves developing xenobots with sensory feedback, allowing them to respond to chemical, thermal, or pressure stimuli. Such sensory xenobots could monitor ecosystems for signs of stress, such as changes in pH levels or temperature fluctuations, and alert researchers to early warning signs of ecological damage. In medical contexts, sensory-enabled xenobots might detect abnormalities within the body, such as high acidity levels around a tumor, and respond by releasing therapeutic compounds or alerting medical staff.

Looking further ahead, xenobots could be designed to interact and communicate with one another. Swarm behavior is common in nature—think of how bees communicate to find food or how schools of fish move together—and similar concepts could be applied to xenobot swarms. Such swarms could work in concert to achieve large-scale tasks, like repairing tissue or removing a significant amount of pollutants from a contaminated site.

Each individual xenobot would play a role within the larger system, contributing to a collective goal, which would make them far more effective than individual units working alone.

Ethical and Philosophical Implications

The idea of programmable living machines, however, raises ethical questions. What does it mean to create organisms for a specific purpose? Unlike traditional machines, xenobots are not entirely artificial; they are living cells programmed to serve a function. This dual nature raises concerns about the boundaries of life and the ethical implications of using biological organisms in a similar way to machines. Are xenobots considered alive? Do they have any form of rights or protection?

Moreover, the potential for xenobots to evolve or mutate—if unintentionally released into the environment—poses risks. Even though current xenobots have a limited lifespan and lack the ability to reproduce, as research advances, new generations of xenobots might have longer lifespans or regenerative abilities that could impact ecosystems. If xenobots were to be accidentally introduced into natural environments, the consequences could be unpredictable, creating ethical questions about how they should be contained and regulated.

Xenobots and the Future of Life Science

The future of xenobots could transform how we understand biology itself. Rather than seeing life as something that evolves purely through natural selection, xenobots introduce a form of directed evolution. They are an example of synthetic life forms created with a specific purpose and potentially altered with each new generation. This technology shifts our view of evolution from a slow, unguided process to one that is intentional and goal-directed, driven by human intervention and creativity.

Xenobots could also change how we think about human-machine interfaces. Unlike traditional robotics, which rely on electrical circuits and rigid frameworks, xenobots represent an entirely organic form of technology.

This raises the possibility of merging organic systems with human tissues, potentially creating new types of implants or devices that integrate seamlessly with our biology. Such developments could redefine how humans interact with technology, creating tools that our bodies accept as naturally as they would any other biological tissue.

Challenges and Future Research

For xenobots to reach their full potential, researchers must overcome significant challenges. One of the biggest hurdles is understanding and refining the control of these biological machines. The cells that make up xenobots have their own natural behaviors and signals, which can interfere with programmed instructions. Overcoming this challenge may require more advanced computational modeling and bioengineering techniques, possibly leading to the development of hybrid xenobots that combine synthetic and natural components.

Another challenge lies in scaling up xenobot production. While scientists have successfully created xenobots in the lab, mass production remains a complex and expensive process. Advances in cell culture techniques, automation, and bioengineering may eventually make it feasible to produce xenobots on a larger scale, bringing their applications closer to real-world scenarios.

A New Frontier for Humanity

The evolution of xenobots marks a new era in technology and biology, merging the organic and synthetic in unprecedented ways.

These living machines are not just tools; they embody a new approach to creating functional organisms from scratch, crafted to perform specific tasks that can transform medicine, environmental science, and beyond.

As we continue to explore and refine xenobots, they will challenge our assumptions about life, blur the boundaries between nature and technology, and redefine our place within both the biological and technological worlds.

The xenobot chronicles are only just beginning, and where they lead could fundamentally alter our relationship with life itself.

References:

Foundational Research on Xenobots

Kriegman, S., Blackiston, D., Levin, M., & Bongard, J. (2020). A scalable pipeline for designing reconfigurable organisms. *Proceedings of the National Academy of Sciences*, 117(4), 1853-1859.

Blackiston, D., Levin, M., & Kriegman, S. (2021). Morphogenesis and autonomous self-repair in programmable biological robots. *Nature Communications*, 12(1), 1-9.

Biology and Technology Merging

Levin, M. (2020). Bioelectric signaling: Reprogrammable circuits underlying embryogenesis, regeneration, and cancer. *Cell*, 184(8), 1971-1984.

Bongard, J., & Levin, M. (2021). Living Things are Information Processors: Biocomputing and Recombinant Bioengineering. *Frontiers in Robotics and AI*, 8, 72.

Neural Networks and AI in Xenobots

Silver, D., et al. (2016). Mastering the game of Go with deep neural networks and tree search. *Nature*, 529(7587), 484–489.

Kriegeskorte, N., & Douglas, P. K. (2018). Cognitive computational neuroscience. *Nature Neuroscience*, 21(9), 1148–1160.

Applications in Neurology and Space Exploration

Parpura, V., et al. (2018). Neuromodulation and future nanoscale therapies for nervous system disorders. *Nature Nanotechnology*, 13(10), 927-940.

Mojarradi, M., & Farmer, J. (2021). Biomimetics and AI: Challenges of deploying bio-robots for planetary exploration. *Astrobiology*, 21(12), 1422-1435.

Regulatory and Ethical Considerations

de Miguel Beriain, I., & Rueda, J. (2021). Ethical and legal challenges in the application of synthetic biology. *Bioethics*, 35(7), 691–704.

Akst, J. (2020). The Need for Regulation in Emerging Biotechnologies. *The Scientist*, 34(3), 18-23.

Interdisciplinary Collaboration and Future Directions

Bruggeman, F. J., & Westerhoff, H. V. (2007). The nature of systems biology. *Trends in Microbiology*, 15(2), 45-50.

Raj, A., & van Oudenaarden, A. (2008). Stochastic gene expression and its consequences. *Cell*, 135(2), 216–226.

Advanced Bioengineering and Xenobots

Ross, M. (2018). Microengineering living tissues: Principles and applications. *Annual Review of Biomedical Engineering*, 20, 303-327.

Cvetkovic, C., et al. (2014). Three-dimensionally printed biological machines powered by skeletal muscle. *Proceedings of the National Academy of Sciences*, 111(28), 10125-10130.

Kim, J., Campbell, A. S., & Wang, J. (2017). Wearable and implantable biosensors. *Chemical Reviews*, 118(14), 6457-6494.

Synthetic Biology and Programmable Life

Basu, S., et al. (2005). A synthetic multicellular system for programmed pattern formation. *Nature*, 434(7037), 1130-1134.

Liu, W., & Zhang, L. (2019). Microbots for environmental and biomedical applications. *Advanced Functional Materials*, 29(26), 1806878.

Wolpe, P. R. (2002). Treatment, enhancement, and the ethics of neurotherapeutics. *Brain and Cognition*, 50(3), 387-395.

Emergent Behaviors and Self-Replication

Gilbert, S. F. (2020). Developmental plasticity and evolutionary theory. *Nature Reviews Genetics*, 21(4), 202-215.

Kriegman, S., et al. (2021). Kinematic self-replication in reconfigurable organisms. *Proceedings of the National Academy of Sciences*, 118(49), e2112672118.

Nanotechnology and Biohybrid Systems

Bhushan, B. (2018). Bioinspired water collection methods for microbots. *Advances in Colloid and Interface Science*, 256, 111-125.

Cui, H., et al. (2021). Nanorobotics in living organisms for medicine. *Nature Biomedical Engineering*, 5(7), 624-635.

Esteban-Fernández de Ávila, B., et al. (2018). Micromotor-enabled active drug delivery for in vivo treatment of stomach infection. *Nature Communications*, 9(1), 1-8.

Cognitive Neuroscience and Computational Biology

Buckner, R. L., & Krienen, F. M. (2013). The evolution of distributed association networks in the human brain. *Trends in Cognitive Sciences*, 17(12), 648-665.

Ashby, W. R. (1960). Design for a brain: The origin of adaptive behavior. *Springer Science & Business Media*.

Wilson, C., et al. (2020). Modeling adaptive neural dynamics using artificial intelligence. *Frontiers in Computational Neuroscience*, 14, 68.

Ethics, Regulation, and Risk Assessment

Greely, H. T. (2019). CRISPR People: The Science and Ethics of Editing Humans. *The MIT Press*.

Kourany, J. (2018). Philosophy of science after feminism: A focus on science ethics. *Oxford University Press*.

Resnik, D. B., & Sharp, R. R. (2006). Protecting third parties in human subjects research. *The Journal of Clinical Ethics*, 17(3), 256-263.

Interdisciplinary Innovation and Collaboration

Friston, K. J., et al. (2017). Active inference: A process theory. *Neural Computation*, 29(1), 1-49.

Searls, D. B. (2010). The roots of bioinformatics in theoretical biology. *PLoS Computational Biology*, 6(3), e1000740.

Johnson, D. R., & Wetmore, J. M. (2008). Engineering ethics for a globalized world. *Technology and Society Magazine*, 27(4), 15-23.

Biological Robotics and Synthetic Systems

Doudna, J. A., & Charpentier, E. (2014). The new frontier of genome engineering with CRISPR-Cas9. *Science*, 346(6213), 1258096.

Sundaram, S., et al. (2021). Soft robotics for chemotactic exploration and actuation. *Science Robotics*, 6(58), eabe6064.

Leduc, C., & Arslan, M. (2018). Biologically inspired locomotion for soft robots. *Advanced Materials*, 30(10), 1707333.

Molecular and Cellular Mechanisms

Albrecht, D. R., et al. (2010). Microfluidics-integrated technologies for exploring intercellular signaling. *Nature Reviews Molecular Cell Biology*, 11(2), 127-138.

Wilson, E. O., & Hölldobler, B. (2005). Eusociality: Origin and consequences. *Proceedings of the National Academy of Sciences*, 102(38), 13367-13371.

Landgraf, M., & Schuman, E. M. (2021). Molecular basis of adaptive plasticity. *Annual Review of Neuroscience*, 44, 1-23.

Emerging Computational Techniques

Mehta, A., & Schwab, D. J. (2020). Emergent dynamics and thermodynamics of computation. *Physical Review Letters*, 125(4), 048101.

Hassabis, D., et al. (2017). Neuroscience-inspired artificial intelligence. *Neuron*, 95(2), 245-258.

Murphy, K. P. (2012). Machine Learning: A Probabilistic Perspective. *MIT Press*.

Microbiological Interfaces and Biohybrid Systems

Hwang, G., et al. (2020). Biohybrid microrobots for microbiota delivery. *Nature Communications*, 11, 5761.

Zhang, L., et al. (2022). Bioengineered bacterial systems for environmental applications. *ACS Nano*, 16(3), 1892-1901.

Sitti, M., et al. (2015). Biologically inspired miniature soft robots. *Nature Reviews Materials*, 1(9), 16008.

Environmental Impacts and Bioenergy

Zhu, B., et al. (2021). Biomimetic design for energy-efficient nanotechnology. *Nano Energy*, 85, 105939.

Boysen, L. R., & Clifford, K. (2018). The role of synthetic biology in achieving sustainability goals. *Nature Sustainability*, 1(10), 556-563.

Nanda, J., & Misra, R. D. K. (2017). Emerging strategies for bioenergy production using hybrid systems. *Bioresource Technology*, 245, 1507-1521.

Ethics, Policy, and Regulation in Synthetic Life

Jasanoff, S. (2004). States of Knowledge: The Co-Production of Science and the Social Order. *Routledge*.

Rabinow, P., & Bennett, G. (2012). Designing human practices: An experiment with synthetic biology. *University of Chicago Press*.

Shapiro, J. A. (2011). Evolution: A View from the 21st Century. *FT Press*.

Collaborative Research and Interdisciplinary Studies

Phillips, R., Kondev, J., & Theriot, J. (2012). Physical Biology of the Cell. *Garland Science*.

Succi, S. (2018). The lattice Boltzmann equation for complex flows. *Oxford University Press*.

Advanced Robotics and Biodesign

Laschi, C., & Cianchetti, M. (2014). Soft robotics: New perspectives for robot bodyware and control. *Frontiers in Bioengineering and Biotechnology*, 2, 3.

Majidi, C. (2019). Soft robotics: A perspective—Current trends and prospects for the future. *Soft Robotics*, 7(2), 131-136.

Tolley, M. T., & Lipson, H. (2016). Additive manufacturing for robust soft robotics. *Advanced Materials*, 28(27), 5700-5712.

Morphogenesis and Tissue Engineering

Takeuchi, S., et al. (2005). Pushing the limits of tissue engineering using microfluidics. *Nature Materials*, 4(6), 525-530.

Khademhosseini, A., & Langer, R. (2016). A decade of progress in tissue engineering. *Nature Protocols*, 11(10), 1775-1781.

Li, Y., et al. (2020). Bioinspired flexible materials: Design and mechanics. *Progress in Materials Science*, 110, 100630.

Synthetic and Programmable Organisms

Muller, P., et al. (2012). Morphogen gradients in development: From form to function. *Annual Review of Cell and Developmental Biology*, 28, 771-798.

Brophy, J. A. N., & Voigt, C. A. (2014). Principles of genetic circuit design. *Nature Methods*, 11(5), 508-520.

Shah, S., & Joshi, D. (2021). CRISPR-enabled development of living programmable systems. *ACS Synthetic Biology*, 10(4), 756-765.

Emerging Nanobiotechnology

Le, T., et al. (2018). Emerging trends in nanobiotechnology for precision medicine. *Nano Today*, 21, 1-14.

Park, B.-W., et al. (2017). Microrobotics: Magnetically powered nanomotors for diagnostics and therapy. *Advanced Materials*, 29(20), 1605682.

Sochol, R. D., et al. (2018). 3D printing with microfluidic technologies. *Science Advances*, 4(8), eaau6989.

Neuroscience and Artificial Intelligence

Rolnick, D., et al. (2019). Tackling climate change with machine learning. *Nature Climate Change*, 9(12), 976-982.

Lillicrap, T. P., et al. (2020). Backpropagation and the brain. *Nature Reviews Neuroscience*, 21(6), 335-346.

Turing, A. M. (1952). The chemical basis of morphogenesis. *Philosophical Transactions of the Royal Society of London. Series B, Biological Sciences*, 237(641), 37-72.

Bioethics and Societal Implications

Stilgoe, J., Owen, R., & Macnaghten, P. (2013). Developing a framework for responsible innovation. *Research Policy*, 42(9), 1568-1580.

Nichols, S. P., & Weldon, B. D. (2009). Interdisciplinary challenges for emerging technologies: Synthetic biology and ethics. *Science and Engineering Ethics*, 15(4), 529-539.

Bedau, M. A., & Triant, M. E. (2011). Emergent behaviors in synthetic life. *BioEssays*, 33(7), 527-531.

Interdisciplinary Collaboration and Challenges

Kitano, H. (2002). Computational systems biology. *Nature*, 420(6912), 206-210.

Taylor, T., & Dorin, A. (2021). Evolutionary models in bio-inspired design: Theory to practice. *Biological Theory*, 16(1), 65-84.

Innovative Techniques in Biodesign and Robotics

Rus, D., & Tolley, M. T. (2015). Design, fabrication, and control of soft robots. *Nature*, 521(7553), 467–475.

Wood, R. J. (2008). The first takeoff of a biologically inspired insect-scale robot. *IEEE Transactions on Robotics*, 24(2), 341–347.

Kamm, R. D., et al. (2018). Perspective: The promise of multi-cellular engineered living systems. *APL Bioengineering*, 2(4), 040901.

Self-Organization and Evolutionary Robotics

Funes, P., & Pollack, J. B. (1998). Evolutionary robotics: Adaptive self-organization of collective behaviors. *Artificial Life*, 4(4), 337–357.

Schrödinger, E. (1944). What is Life? The Physical Aspect of the Living Cell. *Cambridge University Press*.

Bongard, J. (2011). Morphological change in machines accelerates the evolution of robust behavior. *Proceedings of the National Academy of Sciences*, 108(4), 1234–1239.

Mechanics of Biohybrid Systems

Kim, S., Laschi, C., & Trimmer, B. (2013). Soft robotics: A bioinspired evolution in robotics. *Trends in Biotechnology*, 31(5), 287–294.

Richards, J., et al. (2020). Biomechanics of biohybrid robots: Perspectives on design. *Journal of the Royal Society Interface*, 17(164), 20190504.

Xi, W., et al. (2018). Soft micromotors with dynamic structural adaptation for autonomous drug delivery. *Advanced Materials*, 30(50), 1805299.

Artificial Intelligence and Neural Integration

Lake, B. M., et al. (2017). Building machines that learn and think like people. *Behavioral and Brain Sciences*, 40, e253.

Botvinick, M., & Cohen, J. (1998). Rubber hands "feel" touch that eyes see. *Nature*, 391(6669), 756.

Roy, S., et al. (2018). Augmenting neural networks with bio-inspired circuits for cognitive tasks. *Nature Machine Intelligence*, 1, 39–50.

Synthetic Biology and Genetic Programming

Church, G. M., & Regis, E. (2014). Regenesis: How Synthetic Biology Will Reinvent Nature and Ourselves. *Basic Books*.

Gardner, T. S., Cantor, C. R., & Collins, J. J. (2000). Construction of a genetic toggle switch in Escherichia coli. *Nature*, 403(6767), 339–342.

Sole, R. V., & Macia, J. (2013). Synthetic biology: From curiosity to applications. *Philosophical Transactions of the Royal Society B: Biological Sciences*, 368(1625), 20120082.

Sustainability and Energy Applications

Lee, J., et al. (2020). Biodegradable and biocompatible energy storage devices. *Trends in Biotechnology*, 38(5), 465–477.

Su, J., et al. (2021). Harnessing living biohybrid materials for energy production. *ACS Applied Materials & Interfaces*, 13(6), 7431–7442.

Xie, L., et al. (2019). Bio-inspired energy storage and transfer systems. *Nature Reviews Materials*, 4(5), 349–368.

Regulatory Frameworks and Bioethics

Sandler, R., & Basl, J. (2013). Designer Biology: The Ethics of Intensively Engineering Biological and Ecological Systems. *Lexington Books.*

Aicardi, C., et al. (2018). Ethical challenges of emerging robotics and AI technologies. *Ethics and Information Technology*, 20(1), 1–8.

Interdisciplinary Research Success Stories

Alon, U. (2003). Biological networks: The tinkerer as an engineer. *Science*, 301(5641), 1866–1867.

Mittal, S., & Deb, K. (2010). Three-dimensional modeling of interdisciplinary research impact. *Journal of Computational Science*, 1(3), 157–163.

Wiggins, G. A. (2014). Interdisciplinary collaboration in science and technology: A primer. *Innovations in Systems and Software Engineering*, 10(1), 1–10.

Advanced Biological Robotics

Pfeifer, R., & Bongard, J. C. (2007). How the Body Shapes the Way We Think: A New View of Intelligence. *MIT Press.*

Brooks, R. A. (1991). Intelligence without representation. *Artificial Intelligence*, 47(1-3), 139–159.

Shepherd, R. F., et al. (2011). Multigait soft robot. *Proceedings of the National Academy of Sciences*, 108(51), 20400–20403.

Cellular Mechanics and Bioengineering

Discher, D. E., Mooney, D. J., & Zandstra, P. W. (2009). Growth factors, matrices, and forces combine and control stem cells. *Science*, 324(5935), 1673–1677.

Ingber, D. E. (2006). Cellular mechanotransduction: Putting all the pieces together again. *The FASEB Journal*, 20(7), 811–827.

Kamal, J. M., et al. (2019). Computational methods for studying morphogenesis. *Nature Reviews Genetics*, 20(10), 613–627.

Evolutionary Approaches to Robotics and Biology

Floreano, D., & Keller, L. (2010). Evolution of adaptive behaviour in robots by means of Darwinian selection. *PLoS Biology*, 8(1), e1000292.

Bongard, J. C., & Lipson, H. (2007). Automated reverse engineering of nonlinear dynamical systems. *Proceedings of the National Academy of Sciences*, 104(24), 9943–9948.

Wagner, G. P., & Altenberg, L. (1996). Complex adaptations and the evolution of evolvability. *Evolution*, 50(3), 967–976.

Nanobiotechnology and Biohybrid Materials

Balasubramanian, S., et al. (2018). DNA-based molecular machines. *Accounts of Chemical Research*, 51(6), 1550–1561.

Park, J., et al. (2019). Biohybrid soft robots: A marriage of synthetic and living components. *Science Robotics*, 4(30), eaax1985.

Wang, H. F., et al. (2020). Progress in soft robotics with biological sensing and actuation. *Advanced Materials Technologies*, 5(9), 2000371.

Artificial Intelligence in Biological Systems

Schmidhuber, J. (2015). Deep learning in neural networks: An overview. *Neural Networks*, 61, 85–117.

Arulkumaran, K., et al. (2017). Deep reinforcement learning: A brief survey. *IEEE Signal Processing Magazine*, 34(6), 26–38.

Poldrack, R. A. (2006). Can cognitive processes be inferred from neuroimaging data? *Trends in Cognitive Sciences*, 10(2), 59–63.

Synthetic Biology and Living Systems

Bashor, C. J., & Collins, J. J. (2018). Insulating gene circuits from context by RNA processing. *Nature Biotechnology*, 36(7), 593–595.

Cardinale, S., & Arkin, A. P. (2012). Contextualizing synthetic biology. *Nature Reviews Molecular Cell Biology*, 13(9), 614–627.

Cameron, D. E., Bashor, C. J., & Collins, J. J. (2014). A brief history of synthetic biology. *Nature Reviews Microbiology*, 12(5), 381–390.

Ethical and Philosophical Dimensions

Beauchamp, T. L., & Childress, J. F. (2013). Principles of Biomedical Ethics (7th Edition). *Oxford University Press*.

Sandler, R. (2014). Ethics and Emerging Technologies. *Springer*.

Gunkel, D. J. (2018). Robot Rights. *MIT Press*.

Energy and Environmental Implications

Zhang, Z., et al. (2022). Biohybrid systems for sustainable energy applications. *Advanced Energy Materials*, 12(3), 2102100.

Wang, L., et al. (2020). Bio-inspired materials for advanced energy storage systems. *Chemical Society Reviews*, 49(10), 3511–3534.

Shi, Y., et al. (2019). Bio-inspired energy harvesting systems: From materials to devices. *Nano Energy*, 57, 851–871.

Collaborative Interdisciplinary Studies

Krohs, U., & Kroes, P. (2009). Functions in Biological and Artificial Worlds: Comparative Philosophical Perspectives. *MIT Press*.

Mitchell, M. (2009). Complexity: A Guided Tour. *Oxford University Press*.

Nicolis, G., & Prigogine, I. (1977). Self-Organization in Nonequilibrium Systems. *Wiley*.

Cutting-Edge Biocomputing and Cellular Robotics

Cvetkovic, C., et al. (2014). Three-dimensionally printed biological machines powered by skeletal muscle. *Proceedings of the National Academy of Sciences*, 111(28), 10125–10130.

Feinberg, A. W., et al. (2013). Engineered tissue mechanics: A pathway toward building biohybrid robotics. *Annual Review of Biomedical Engineering*, 15, 327–353.

Nawroth, J. C., et al. (2012). A tissue-engineered jellyfish with biomimetic propulsion. *Nature Biotechnology*, 30(8), 792–797.

Dynamic Systems and Self-Healing Materials

Toohey, K. S., et al. (2007). Self-healing materials with microvascular networks. *Nature Materials*, 6(8), 581–585.

Lazarus, N., & Lewis, J. A. (2021). Emerging directions in self-healing structural composites. *Science Advances*, 7(44), eabh2854.

Majidi, C. (2014). Soft-matter engineering for soft robotics. *Advanced Materials*, 26(36), 6410–6421.

Bioinspired Design Principles

Vincent, J. F. V., et al. (2006). Biomimetics: Its practice and theory. *Journal of the Royal Society Interface*, 3(9), 471–482.

Fratzl, P., & Barth, F. G. (2009). Biomaterial systems for mechanosensing and actuation. *Nature*, 462(7272), 442–448.

Bhushan, B. (2009). Biomimetics: Lessons from nature—An overview. *Philosophical Transactions of the Royal Society A: Mathematical, Physical and Engineering Sciences*, 367(1893), 1445–1486.

Morphogenesis and Adaptive Systems

Newman, S. A., & Comper, W. D. (1990). 'Generic' physical mechanisms of morphogenesis and pattern formation. *Development*, 110(1), 1–18.

Edelman, G. M. (1987). Neural Darwinism: The Theory of Neuronal Group Selection. *Basic Books*.

Basanta, D., et al. (2008). Investigating prostate cancer tumour–stroma interactions: Clinical and biological insights from an evolutionary game. *British Journal of Cancer*, 98(6), 1134–1141.

Hybrid Robotics and Modular Design

Paik, J., et al. (2015). Design and control of a bio-inspired multi-degree-of-freedom soft robot. *Bioinspiration & Biomimetics*, 10(2), 025001.

Laschi, C., et al. (2016). Design and development of soft robotics: A focus on biohybrid soft robots. *Soft Robotics*, 3(4), 204–217.

Majidi, C. (2018). Flexible circuits and soft robotics for adaptive materials and morphing structures. *Accounts of Chemical Research*, 51(4), 723–732.

Synthetic Biology for Programmable Systems

Nielsen, A. A. K., et al. (2016). Genetic circuit design automation. *Science*, 352(6281), aac7341.

Purnick, P. E., & Weiss, R. (2009). The second wave of synthetic biology: From modules to systems. *Nature Reviews Molecular Cell Biology*, 10(6), 410–422.

Smolke, C. D. (2009). Building outside of the box: iGEM and the BioBricks Foundation. *Nature Biotechnology*, 27(12), 1099–1102.

Nanotechnology in Biological Integration

Peppas, N. A., et al. (2006). Hydrogels in biology and medicine: From molecular principles to bionanotechnology. *Advanced Materials*, 18(11), 1345–1360.

Kotov, N. A. (2006). Inorganic nanoparticles as protein mimics. *Science*, 314(5802), 604–607.

Nguyen, V. H., et al. (2018). Soft nanotechnology and its implications for future therapeutic designs. *Nanomedicine: Nanotechnology, Biology and Medicine*, 14(3), 945–957.

Ethics, Policy, and Regulation

Resnik, D. B., & Tinkle, S. S. (2007). Ethical issues in nanomedicine. *Nanomedicine: Nanotechnology, Biology, and Medicine*, 3(4), 345–350.

Kahan, D. M., et al. (2012). The tragedy of the risk-perception commons: Culture conflict, rationality conflict, and climate change. *Temple University Legal Studies Research Paper*.

Renn, O., & Roco, M. C. (2006). Nanotechnology and the need for risk governance. *Journal of Nanoparticle Research*, 8(2), 153–191.

Future Directions in Xenobot Research

Kriegman, S., et al. (2021). Scalable simulation-based evolutionary design of multicellular living systems. *PNAS*, 118(15), e2017217118.

Tewksbury, J. J., et al. (2014). Natural history's place in science and society. *BioScience*, 64(4), 300–310.

Moulton, D. E., et al. (2020). Beyond Turing: Morphogenesis in heterogeneous reaction-diffusion-mechanics systems. *Physics of Life Reviews*, 34, 118–133.

Advanced Biological Modeling and Computational Simulations

Couzin, I. D., et al. (2005). Effective leadership and decision-making in animal groups on the move. *Nature*, 433(7025), 513–516.

Keller, E. F. (2002). Making Sense of Life: Explaining Biological Development with Models, Metaphors, and Machines. *Harvard University Press*.

Morris, S. C. (2003). Life's Solution: Inevitable Humans in a Lonely Universe. *Cambridge University Press*.

Emergent Behaviors and Collective Intelligence

Gazi, V., & Passino, K. M. (2004). Stability analysis of social foraging swarms. *IEEE Transactions on Systems, Man, and Cybernetics-Part B: Cybernetics*, 34(1), 539–557.

Garnier, S., Gautrais, J., & Theraulaz, G. (2007). The biological principles of swarm intelligence. *Swarm Intelligence*, 1(1), 3–31.

Rubenstein, M., et al. (2014). Programmable self-assembly in a thousand-robot swarm. *Science*, 345(6198), 795–799.

Bioactuation and Biomechanical Systems

Zhao, X., et al. (2011). Harnessing deformation to control soft, adaptive materials. *Nature Materials*, 10(8), 606–617.

Nawroth, J. C., et al. (2012). Muscle-powered biomimetic jellyfish for artificial propulsion. *Nature Biotechnology*, 30(8), 792–797.

Richards, J., et al. (2021). Responsive hydrogels for dynamic bioactuation. *Nature Reviews Materials*, 6(8), 616–630.

Regenerative Biology and Tissue Dynamics

McCusker, C., Bryant, S. V., & Gardiner, D. M. (2015). The axolotl limb regeneration model as a guide for regenerative medicine. *Regeneration*, 2(2), 54–71.

Poss, K. D. (2010). Advances in understanding tissue regenerative capacity and mechanisms in animals. *Nature Reviews Genetics*, 11(10), 710–722.

Lobo, D., et al. (2014). Model-based discovery of molecular regulators controlling dynamic cellular processes. *PLoS Computational Biology*, 10(5), e1003638.

Self-Repairing Robotics and Material Science

White, S. R., et al. (2001). Autonomic healing of polymer composites. *Nature*, 409(6822), 794–797.

Zhang, Z., & Yin, Y. (2019). Emerging soft robotics: Materials, actuation, and sensors. *Advanced Functional Materials*, 29(22), 1806692.

Sitti, M., et al. (2015). Biomedical applications of untethered mobile milli/microrobots. *Proceedings of the IEEE*, 103(2), 205–224.

Synthetic Intelligence and Bio-Cognition

Clark, A. (1997). Being There: Putting Brain, Body, and World Together Again. *MIT Press*.

Tononi, G., & Edelman, G. M. (1998). Consciousness and complexity. *Science*, 282(5395), 1846–1851.

Friston, K. (2010). The free-energy principle: A unified brain theory? *Nature Reviews Neuroscience*, 11(2), 127–138.

Ecological Applications of Biohybrid Systems

Komeili, A., et al. (2006). Magnetosomes are cell membrane invaginations organized by the actin-like protein MamK. *Science*, 311(5758), 242–245.

Armstrong, C. M., & Winey, M. (2009). Cytoskeletal proteins in cellular structure and adaptability. *Nature Cell Biology*, 11(1), 3–8.

Tian, X., et al. (2017). Eco-engineering of biohybrid organisms for waste management. *Trends in Biotechnology*, 35(4), 328–340.

Ethics and Philosophy of Machine-Life Hybrids

Bostrom, N. (2003). Ethical issues in advanced artificial intelligence. *Cognitive, Emotive and Ethical Aspects of Decision Making in Humans and in Artificial Intelligence*, 1(2), 12–17.

Floridi, L., & Sanders, J. W. (2004). On the morality of artificial agents. *Minds and Machines*, 14(3), 349–379.

Sparrow, R. (2007). Killer robots. *Journal of Applied Philosophy*, 24(1), 62–77.

Frontiers in Biotechnological Applications

Kasper, D. L., & Ausubel, F. M. (2010). Advances in microbial biotechnology: Implications for the future of healthcare. *Nature Reviews Microbiology*, 8(10), 767–778.

Tanaka, K., et al. (2021). Biohybrid microbots with living cell actuators. *ACS Nano*, 15(9), 14467–14476.

Mirkin, C. A., & Niemeyer, C. M. (2007). Nanobiotechnology: Concepts, Applications and Perspectives. *Wiley-VCH*.

www.ingramcontent.com/pod-product-compliance
Lightning Source LLC
Chambersburg PA
CBHW071016240526
45469CB00006BD/1944